AF696651

# STEREOTAXIC BRAIN ATLAS OF THE EGYPTIAN FRUIT BAT

# STEREOTAXIC BRAIN ATLAS OF THE EGYPTIAN FRUIT BAT


Raya Eilam-Altstädter
*Weizmann Institute of Science, Rehovot, Israel*

Liora Las
*Weizmann Institute of Science, Rehovot, Israel*

Menno P. Witter
*Norwegian University of Science and Technology, Trondheim, Norway*

Nachum Ulanovsky
*Weizmann Institute of Science, Rehovot, Israel*



ELSEVIER

ACADEMIC PRESS
An imprint of Elsevier

Academic Press is an imprint of Elsevier
125 London Wall, London EC2Y 5AS, United Kingdom
525 B Street, Suite 1650, San Diego, CA 92101, United States
50 Hampshire Street, 5th Floor, Cambridge, MA 02139, United States
The Boulevard, Langford Lane, Kidlington, Oxford OX5 1GB, United Kingdom

Copyright © 2021 Elsevier Inc. All rights reserved.

No part of this publication may be reproduced or transmitted in any form or by any means, electronic or mechanical, including photocopying, recording, or any information storage and retrieval system, without permission in writing from the publisher. Details on how to seek permission, further information about the Publisher's permissions policies and our arrangements with organizations such as the Copyright Clearance Center and the Copyright Licensing Agency, can be found at our website: www.elsevier.com/permissions.

This book and the individual contributions contained in it are protected under copyright by the Publisher (other than as may be noted herein).

**Notices**
Knowledge and best practice in this field are constantly changing. As new research and experience broaden our understanding, changes in research methods, professional practices, or medical treatment may become necessary.

Practitioners and researchers must always rely on their own experience and knowledge in evaluating and using any information, methods, compounds, or experiments described herein. In using such information or methods they should be mindful of their own safety and the safety of others, including parties for whom they have a professional responsibility.

To the fullest extent of the law, neither the Publisher nor the authors, contributors, or editors, assume any liability for any injury and/or damage to persons or property as a matter of products liability, negligence or otherwise, or from any use or operation of any methods, products, instructions, or ideas contained in the material herein.

**Library of Congress Cataloging-in-Publication Data**
A catalog record for this book is available from the Library of Congress

**British Library Cataloguing-in-Publication Data**
A catalogue record for this book is available from the British Library

ISBN 978-0-12-819297-9

For information on all Academic Press publications visit our website at https://www.elsevier.com/books-and-journals

*Cover Design:* Noa David Geller
*Atlas Plates Graphical Design:* Noa David Geller

*Publisher:* Nikki Levy
*Acquisitions Editor:* Natalie Farra
*Editorial Project Manager:* Sam W. Young
*Production Project Manager:* Niranjan Bhaskaran
*Designer:* Christian J. Bilbow

Typeset by SPi Global, India

ELSEVIER Book Aid International
Working together to grow libraries in developing countries
www.elsevier.com • www.bookaid.org

# Dedication

R. E.-A. dedicates this book to the memory of her beloved parents Lea and Walter Altstädter

# Contents



For additional information on the topics covered in the book, visit the companion site: https://www.elsevier.com/books-and-journals/book-companion/9780128192979

# Acknowledgments

We thank Eva Himovichi for help in initial histological processing and photography, and Mariana Melcón for initial rotation and alignment of the plates. Wairimu Gatome performed most of the tract-tracing injections in the hippocampal formation, on a large group of 17 bats, which served as a basis for delineating the hippocampal formation substructures; Heidi Kleven analyzed these track-tracing data and contributed to the delineations. Bente Jacobsen contributed to the preparation of the Appendixes. Saikat Ray performed tract-tracing injections in three bats and also stained prefrontal cortex sections with Nissl, AChE, WFS1, and NNat, and used these to delineate prefrontal subfields—cingulate, prelimbic, and infralimbic cortices; Efrat Solomon assisted him with processing these brains. We thank Inbal Biton for high-resolution MRI images taken from a separate individual bat, which aided in 3D reconstructions of the hippocampal formation (Appendix 6). We thank Michael Okun, Yonatan Katz, and Ilan Lampl for electrophysiological recordings of neuronal responses to oriented bars in the visual cortex, which helped localizing V1. Many thanks to Rena Orman, Leah Krubitzer, Paul Manger, Jeff Wenstrup, Susanne Sterbing-D'Angelo, Kathy Rockland, Michael Yartsev, Saikat Ray, Patrick Hof, Laszlo Zaborsky, Trygve Leergaard, and Shir Maimon for comments and corrections of the atlas. Finally, and most importantly, our utmost thanks go to Noa David Geller from the Graphics Department of the Weizmann Institute of Science, who meticulously carried to perfection all the graphical arrangements and corrections, and also designed the cover image of the atlas—Thank you, Noa!

# Methods

This stereotaxic brain atlas is based on the right hemisphere of a single fresh brain from a female Egyptian fruit bat (*Rousettus aegyptiacus*) that was wild-caught as an adult in Israel. The animal's weight was 143.3 g. Coronal sections were cut perpendicularly to the line that joins lambda and bregma, and were stained for Nissl, acetylcholine esterase (AChE), and the calcium-binding protein Parvalbumin. This atlas provides a major improvement over the previously existing brain atlas of this species (Schneider, 1966): the current atlas is stereotaxic; it includes multiple stainings—Nissl, AChE, and Parvalbumin in the main stereotaxic plates, as well as several other stainings in the appendixes (nonstereotaxic); and it features detailed delineations of major brain structures.

## Stereotaxic surgery

We placed the anesthetized bat in a Kopf small-animal stereotact, and adjusted the mouth-bar until lambda and bregma were at the same height, i.e., the head was in a flat-skull position, which we considered as the horizontal plane in the atlas. Bregma was defined as the meeting point of the anterior coronal suture and sagittal suture (Fig. 1, right). Lambda was defined as the meeting point of the lambdoid suture and sagittal suture. The lambdoid suture is quite difficult to visualize in vivo in bats, and therefore we applied $H_2O_2$ to the skull during the surgery, in order to better visualize it, so as to achieve maximal accuracy of the flat-skull horizontal position.

To confirm the correct horizontal positioning of the bat's skull during surgery and the correct coronal cutting of the sections, we made four vertical needle tracks stereotaxically in the horizontal (flat-skull) position. The needles (stainless-steel insect pins) were inserted into the brain's right hemisphere through small craniotomies: they were inserted in a vertical D-V (dorso-ventral) direction—perpendicular to the horizontal plane, and were removed after a few minutes. The four needles were coated with DiI fluorescent dye, diluted in 70% ethanol, which allowed later visualization of the needle tracks in a fluorescent microscope (Fig. 2). The four needles were inserted at the following coordinates (with slight adjustments): (8.5 mm A-P, 1.0 mm M-L), (8.5 mm A-P, 3.5 mm M-L), (5.2 mm A-P, 1.0 mm M-L), and (5.2 mm A-P, 3.5 mm M-L). These A-P (antero-posterior) coordinates were measured from the interaural line (the earbars), and the M-L (medio-lateral) coordinates were measured from the midline. The tracks of these four pins are partially visible in atlas plates 50, 51, 53, 72, 73. All the pins were inserted to approximately 3 mm below the brain surface, and created nice long vertical tracks in the coronal sections taken from this brain (Fig. 2C: see the long red fluorescent DiI track; and see "Histological methods" section). Thus these reference tracks confirmed the correct horizontal (flat-skull) positioning of the bat's head during surgery and the correct coronal orientation of the sections cut for this atlas.

Following pin-insertion surgery, the bat was euthanized with pental. It was not perfused (neither with PBS nor with fixative): instead, the fresh brain was carefully extracted without using any fixative. The extracted brain was then postfixated at 4°C overnight in 2.5% paraformaldehyde, and then for an additional 1 day in 1% paraformaldehyde. The rationale for avoiding perfusion and minimizing the use of fixative is that it prevents brain shrinkage, which is important when making an atlas, because it allows maintaining the stereotaxic precision of the atlas. This has relatively little cost: the blood that remains in the brain due to the lack of perfusion does not affect the quality of Nissl and AChE staining, but it did create a constant background in the Parvalbumin staining, somewhat degrading the quality of Parvalbumin plates. The brain was then put overnight in 30% sucrose at 4°C and was subsequently submerged in 9% gelatin at 37°C for 1 day (we used a low-melting gelatin [Sigma] that is needed to prevent disintegration of the posterior parts of the cortex during cutting; the heating to 37°C is needed to attach the gelatin well to the brain tissue). The brain was then put again overnight in 30% sucrose

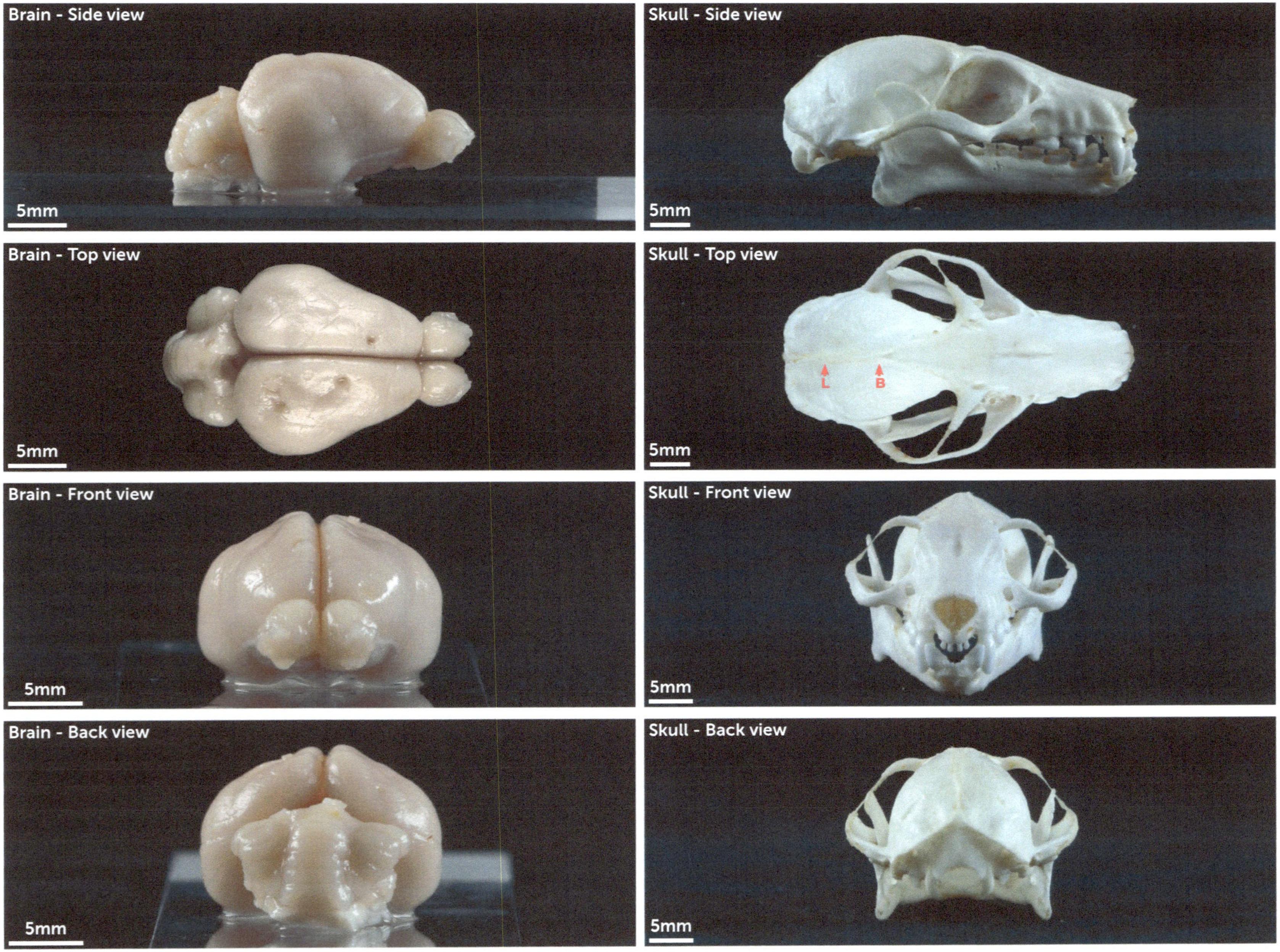


FIG. 1 Photographs of a skull and brain of an Egyptian fruit bat (*Rousettus aegyptiacus*). These photographs were taken from different individuals than the one used for preparing the atlas. The holes in the top view of the brain reflect damage due to surgery. "L" and "B" on the skull mark the lambda and bregma points.

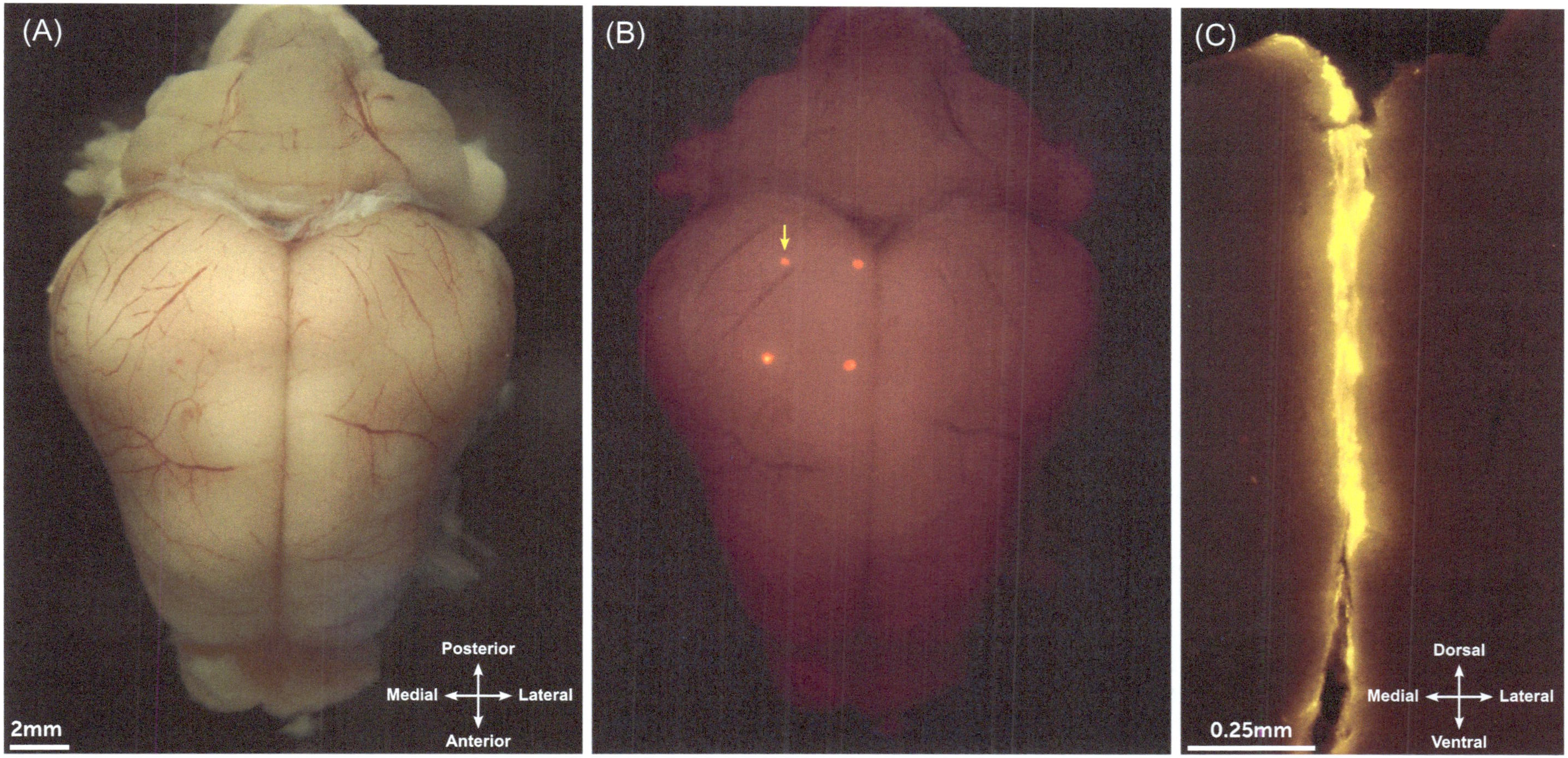


FIG. 2 The brain of the bat that was used for making this atlas. (A, B) Whole-brain photographs: brightfield photograph (A) and fluorescent photograph (B). These photographs were taken immediately after the brain extraction. The four bright *red dots* in B correspond to the four vertical needle tracks coated with DiI (a red-colored fluorescent dye): These four needle tracks were made during surgery as fiduciary markers, in order to calibrate the stereotaxic coordinates and the thickness of the coronal sections, and to verify the flat-skull orientation and the correct coronal cutting of the sections. (C) One of the DiI-coated needle tracks (marked by *yellow arrow* in panel B), taken from a coronal section of the brain. The long straight track (seen as a fluorescent red/orange trace) confirms the correct flat-skull alignment during surgery, and also confirms the correct stereotaxic orientation of the coronal sectioning for the atlas.

at 4°C and was then frozen on a freezing microtome. Floating sections were cut at 41 μm thickness, while the left hemisphere was marked during cutting in order to allow correct left/right orientation when mounting the sections (see later). This average 41-μm thickness was verified by counting the number of sections between the four calibration needle tracks (Fig. 2B and C) and comparing to the stereotaxic distance between the pins as measured in vivo during the surgery. Sections were collected in two series: Nissl (series 1) → AChE/Parvalbumin (series 2). We placed every second Nissl section and every second AChE/Parvalbumin section in the atlas: Thus the interval between consecutive atlas plates (pages) is generally

4 × 41 μm = 164 μm. On each plate of the atlas we show the Nissl section from the Nissl series (series 1) on the left side of the plate, and either AChE or Parvalbumin staining from the AChE/Parvalbumin series (series 2) on the right side of the plate (some of the sections in series 2 were stained for AChE and some for Parvalbumin). All sections, both Nissl and AChE/Parvalbumin, were taken from the right hemisphere (see later). In cases of a damaged section, we took one of the adjacent sections, and the A-P coordinates listed in the atlas were corrected accordingly. Additional exceptions to the 164-μm interplate interval include plates 42–53, containing the dorsal hippocampus; plates 59–63, containing the ventral hippocampus; and plates 76–87, containing the entorhinal cortex—brain areas of particular interest to our laboratories—for which we used half the average interval between plates: 82 μm (i.e., included every Nissl section instead of every second Nissl section). All the coordinates written on the atlas plates correspond to the *actual* sections used.

During cutting on the microtome, we lost the most posterior part of the brain—containing the cerebellum and brainstem, and the most anterior part—containing the olfactory bulb. The atlas thus does not contain the cerebellum, brainstem, and olfactory bulb in stereotaxic coordinates. Nevertheless, for completeness, we took photographs of the cerebellum, brainstem, and olfactory bulb from the same brain, and have placed these illustrative (nonstereotaxic) photographs in Appendix 1.

## Histological methods

We used the right hemisphere for taking the pictures of the atlas, both for the Nissl and for the AChE/Parvalbumin halves of each plate; the Nissl photographs were flipped left-right, so it appears in the atlas as if the Nissl comes from the left hemisphere, but in fact the Nissl are flipped photographs of the right hemisphere. To maintain correct left-right identification during mounting of the floating sections onto the glass slides, we marked the left hemisphere during cutting on the freezing microtome: This marking was done by passing a 21-ga blunt needle horizontally in the A-P direction through the left hemisphere (nondisplayed hemisphere). In one case (plate 79: Parvalbumin), the right-hemispheric section and its neighbors were damaged, so we took the anatomically corresponding left-hemispheric section.

### Cresyl violet Nissl staining

Cresyl violet stains the so-called Nissl substance, which is the rough endoplasmic reticulum of all cells, including both neurons and glia. Sections were mounted from PBS onto Superfrost Plus slides (Menzel-Gläser), three sections on each slide, and were dehydrated for 2 days at room temperature. Slides were then immersed shortly in PBS and were then dehydrated by immersing in increased ethanol concentrations (10 min in each concentration): 50%, 70%, 95% (twice), and 100% (three times). The slides were then dipped in chloroform for 1 h (to remove lipids). Slides were rehydrated with decreased ethanol concentrations: 100% (three times), 95% (twice), 70%, and subsequently dipped for 10–20 min in 0.25% cresyl violet. Slides were immersed shortly in PBS followed by fast wash in 25% ethanol, and differentiated in 50% ethanol + 0.5% acetic acid for about 3 min to reduce nonspecific and Glia staining. Slides were then dehydrated again through increased concentration of ethanol between 70% and 100%. They were put in xylene (10 min, three times), embedded in Entellan (Merck), and covered with a glass coverslip.

### AChE immunohistochemistry

Sections were incubated free-floating for 15 h in 100 mL of stock solution (see later), to which we added 115.6 mg of *S*-acetylthiocholine iodide and 6.0 mg ethopropazine (Sigma). The sections were rinsed with tap water and developed for 2–5 min in 10% K-Ferricyanide. They were then washed in PBS, rinsed with water, and mounted on Superfrost Plus slides, and allowed to dry for 1–2 days at room temperature. Subsequently, they were dehydrated for 10 min in 70%, 95%, and 100% ethanol, then immersed in xylene, embedded in Entellan, and covered with a glass coverslip.

The stock solution for the AChE staining was comprised of three different solutions:

1. 0.5 M sodium acetate buffer (titrated with glacial acid) at pH 5.0.
2. 0.1 M glycine (1.875 g in 250 mL $ddH_2O$).
3. 2 mM $CuSO_4 \cdot 5H_2O$ (1.25 g in 250 mL $ddH_2O$).

These three stock solutions were kept at 4°C. To prepare working solution, we mixed 10 mL of each of the three solutions, added the *S*-acetylthiocholine iodide and ethopropazine mentioned before, and filled the mixed solutions with $ddH_2O$ up to 100 mL.

## Parvalbumin and TH immunohistochemistry

Staining of free-floating sections for Parvalbumin (PV) and Tyrosine Hydroxylase (TH) immunohistochemistry was done as follows (PV—see atlas plates; TH—see Appendix 2). Sections were incubated with hydrogen peroxidase quenching reagent for 30 min (3% $H_2O_2$, 1% HCl in 50% methanol diluted with $ddH_2O$) to quench endogenous hydrogen peroxidase. Next, sections were washed with PBS 1 × and incubated for 90 min in blocking buffer (20% Normal Horse Serum and 2% Triton × 100 in PBS 1 × solution). At the end of the incubation the buffer was sucked out, and the sections were incubated overnight with the primary antibodies anti-PV (1:100, Swant) or anti-TH (1:500, Millipore). On the next day, after washing with PBS 1 ×, the sections were incubated with a biotinylated secondary antibody—donkey antirabbit IgG (1:100, Jackson) for 90 min at room temperature, followed by incubation with avidin-biotin-peroxidase complex (elite ABC kit, Vector). Sections were then washed and incubated with chromogenic substrate 3,3′-diaminobenzidine (DAB, Sigma). Sections were washed and mounted on charged slides (Superfrost Plus), air-dried for 1 day, dehydrated through gradient percentage of ethanols, dipped in xylene solution three times, and were then covered with a glass coverslip with Entellan mounting medium (Merck).

## Oxytocin immunohistochemistry

Staining of free-floating sections for oxytocin immunohistochemistry was done as follows (see Appendix 4). Sections were incubated for 90 min in blocking buffer (20% Normal Horse Serum and 2% Triton × 100 in PBS 1 × solution). Next, the blocking solution was sucked out and the sections were incubated overnight with the primary antibody for neurophysin (mouse, 1:500, Millipore). On the next day, after washing with PBS, the sections were incubated for 90 min at room temperature with goat antimouse secondary antibody (Alexa Fluor 488 conjugated, 1:100, Molecular Probes). Sections were then washed, mounted on charged slides (Superfrost Plus), air-dried for 15 min, and then covered with a glass coverslip with aqua-poly mount medium (Polysciences).

## Double staining

Double staining of free-floating sections for TH, PV, and Choline Acetyl Transferase (ChAT) was done as follows. Fluorescent detection: sections were incubated for 90 min in blocking buffer (20% Normal Horse Serum and 2% Triton × 100 in PBS 1 × solution). Next, the blocking solution was sucked out and the sections were incubated overnight with two primary antibodies: ChAT (goat) and TH (rabbit) antibodies (1:100 and 1:500, respectively, Millipore); or ChAT (goat, Millipore) and PV (rabbit, Swant) antibodies (1:100 and 1:50, respectively). On the next day, after washing with PBS, the sections were incubated for 90 min at room temperature with biotinylated donkey antigoat IgG secondary antibody (1:100, Jackson), to enhance ChAT staining, followed by incubation with an additional secondary antibody, Cy2 conjugated rabbit IgG together with Cy3 conjugated Streptavidin. Sections were then washed, mounted on charged slides (Superfrost Plus), air-dried for 15 min, and then covered with a glass coverslip with aqua-poly mount medium (Polysciences). ChAT and TH double staining is shown in Appendix 3.

## Injections of tracers into the auditory thalamus and auditory cortex

To identify the primary auditory cortex (Au1) and ventral auditory field (AuV), we performed the following tract-tracing injections in auditory thalamus and auditory cortex, in 13 animals (see examples in Appendix 5). Injection coordinates in some of the surgeries (in both areas) were aided by prior electrophysiology recordings that were done immediately before the injections, which identified auditory neuronal responses. In auditory thalamus (MGB) the approximate coordinates were between 2.7 and 3.2 mm anterior to the sinus and 3.8–4.2 lateral to the midline, at roughly 6.0–6.5 mm depth from the brain surface. We injected BDA (BDA-10,000, Molecular Probes; diluted 1:20 in phosphate buffer; iontophoretic injection). In the auditory cortex we conducted both injections of BDA and cholera toxin (cholera toxin subunit B, Alexa Fluor 555 conjugated, Molecular Probes; diluted 1:30 in PBS; pressure-injected using Nanoject, volume 28 nL). Injections in auditory cortex were performed over a wide range of anterior-posterior coordinates, at roughly 7 mm lateral from midline; see Appendix 5 for images.

## Photography of sections

Nissl and AChE photographs for the atlas plates were taken through a Nikon E800 microscope equipped with 0.5 × Plan Apo objective connected to a camera (DS-Qi1, Nikon). Most of the other photographs of sections were taken using a 3DHistech scanner at 20 × magnification—this includes the following: Parvalbumin photographs for the atlas plates, a few of the Nissl plates in the atlas (retaken to improve quality), photographs of the TH staining (Appendix 2) and oxytocin staining (Appendix 4), and the ChAT and TH double-staining photographs (Appendix 3). The photographs of BDA and cholera toxin tracers (Appendix 5) were taken by a fluorescent stereomicroscope (Leica M165 FC) with 12.5:1 zoom ratio with 1 × objective, equipped with a darkfield condenser; the microscope was connected to a monochrome camera (Leica DFC365 FX).

## Stereotaxic reference system

The stereotaxic reference system follows the convention of the Paxinos and Watson rat brain atlas (Paxinos and Watson, 2007), and is based on the horizontal flat-skull position, in which bregma and lambda lie in the same horizontal plane, thus having a zero vertical difference.

The A-P distance between the interaural line and the lambda and bregma point was measured in this bat and in several other bats, and was found to be highly variable as these are wild animals and not a laboratory strain; therefore, we do not report here craniometric measurements of the skull. To achieve more consistent A-P measurements when using this atlas, we transformed the A-P interaural coordinates measured here to A-P "sinus" coordinates, measured from the anterior edge of the posterior transverse sinus that runs between the cortex and cerebellum—the A-P position of this sinus was found to be much more consistent from animal to animal than those of lambda or bregma. This sinus was previously found by one of us (Menno Witter) to be the most reliable anchor for measuring A-P coordinates in rats, especially for the posterior half of the brain. It is therefore highly recommended for the users of this atlas to use the "sinus" A-P coordinates indicated on each atlas plate; these coordinates always correspond to the A-P coordinate of the Nissl half.

The transformation from interaural A-P coordinates to sinus A-P coordinates was done as follows. We performed many stereotaxic calibration surgeries over the years, in which we did electrolytic lesions or injected tracers into dorsal hippocampal area CA1, ventral hippocampus, and infralimbic and prelimbic cortex (the majority of the calibration surgeries were done in dorsal hippocampus). All these lesions and injections were done with bats in the horizontal flat-skull position, using the sinus A-P coordinates. We then found the interaural A-P coordinates at which the lesions or injections reached the target area, by comparing coronal photographs of the injected bats to the atlas plates. This allowed us to determine that 0 mm A-P in sinus coordinates correspond to 2.7 mm A-P in interaural coordinates. We listed the sinus coordinates on each Nissl plate, which corresponds to the interaural coordinate of that plate minus 2.7 mm. As emphasized before, we highly recommend always using the sinus A-P coordinates when working with this atlas—even when targeting anterior brain regions—because the sinus is much more reliable from animal to animal than either the interaural line or bregma.

## Zero coordinates of the atlas in the three planes

*M-L:* Photographed coronal sections were rotated such that the midline is precisely vertical, and were shifted to bring the midline to 0 mm M-L. The zero M-L coordinate thus corresponds to the midline of each section. Note that we always show a bit of the contralateral hemisphere for both the Nissl and AChE/Parvalbumin halves of each plate, thus there are two different 0-ML coordinates on each atlas plate: the users are therefore advised to pay attention to the top scale above each plate, which shows these two 0-ML coordinates. Note that the vertical line in the middle of each plate is *not* the 0-ML coordinate.

*A-P:* The A-P "sinus" coordinate of each plate corresponds to the distance of the Nissl section from the anterior edge of the posterior transverse sinus that runs between the cortex and cerebellum (measured always at 3.3 mm M-L). The zero A-P coordinate lies 0.71 mm posteriorly to the most posterior plate of the atlas (plate 87). It is important to emphasize that since all the atlas A-P coordinates were measured from the sinus at 3.3 mm M-L laterality, it follows that all experimental work that will be based on this atlas must also be done with A-P coordinates measured from the sinus at the same 3.3 mm M-L laterality.

*D-V:* To determine the correct vertical D-V shift of each atlas plate, we performed a stereotaxic surgery on a second bat (weight 157 g, male), which we placed in the Kopf stereotact in a horizontal flat-skull position, and we then marked the lambda and bregma points in the skull by drilling narrow transverse craniotomies (oriented in the M-L direction) at the A-P coordinates of lambda and bregma. We then euthanized the bat and, without any perfusion, we decapitated it, cleaned the skin and muscle tissues, and then we dipped its head—skull and brain together—in 12.5% EDTA/PBS solution (pH 7.4) for 2 weeks, using fresh solution every day. The EDTA solution decalcifies the bone and softens the skull, allowing us to section the skull and brain together. We sectioned the skull and brain parasagittally, using a razor, at 1.8 mm lateral to the midsagittal plane. We then photographed this section, and rotated this photograph to bring the craniotomies drilled in the lambda and bregma to be at the same horizontal position, i.e., we rotated the photograph to a horizontal flat-skull position: Fig. 3A shows this rotated section in the horizontal position. We then used a sagittal Nissl-stained brain section from a third bat, taken at 1.6 mm lateral to midline (Fig. 3B; we note that the 1.6 mm laterality is after correcting for shrinkage, as this third bat was perfused with paraformaldehyde): We rotated it to match the brain rotation shown in Fig. 3A and then we measured in Matlab the cortical outline of this section, after normalizing its A-P extent to the extent of the cortex plates in the atlas: see the red outline in Fig. 3B. This allowed us then to correct the vertical D-V shift of each atlas plate, by shifting the coronal section on the atlas plate up or down, such that the cortex height at 1.6 mm lateral to midline would have the same height as determined based on this measured outline of the sagittal section (Fig. 3B).

At the very posterior parts of the atlas (plates 81–87), where the cortex D-V height changes very rapidly from one coronal section to the next, we adjusted the D-V height of the sections based on the aqueduct of Sylvius, which has an angle of 26 degrees with respect to the horizontal flat-skull line (the 26-degree angle was measured in a sagittally cut brain of a fourth bat).

The zero D-V coordinate of the atlas was set to be the uppermost point of the *skull* at 1.6-mm M-L; in other words, 0 mm D-V in the atlas denotes the height of the top of the skull at 1.6 mm lateral to midline. This occurs around the A-P position of the dorsal hippocampus, in plates 44–49 (see Fig. 3B, red arrow). The skull thickness at that location was determined using a caliper, by averaging over the thicknesses of seven skull pieces that were chipped approximately above the dorsal hippocampus in one of the bats; the mean skull thickness at that location was measured to be 0.45 mm, hence the

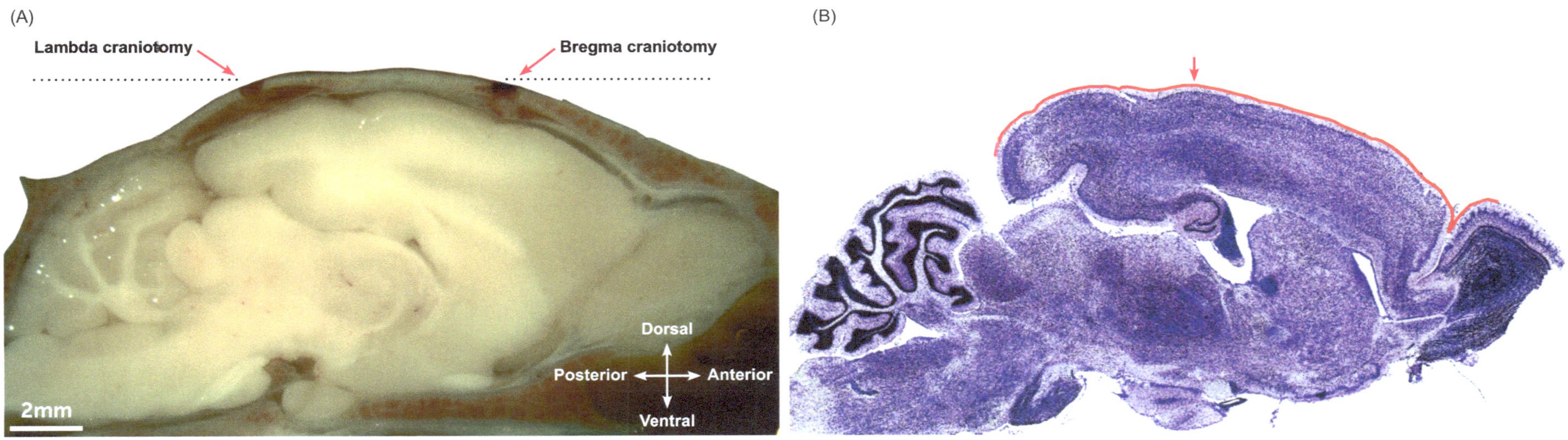


FIG. 3 Determining the vertical shift of each atlas plate. To this end, we used two bats (bat #2 and bat #3) in addition to the atlas bat (bat #1). (A) Sagittal section (taken 1.8 mm lateral to midline) from bat #2, which was cut through brain and skull together in this bat; this sagittal image was rotated to bring it to the horizontal flat-skull position, such that the bregma and lambda are at the same horizontal plane (*dotted line*). This bat was euthanized without perfusion, then underwent decalcification using EDTH, which allowed cutting through the skull and brain together. Indicated are the craniotomies that we drilled in the bregma and lambda points while this bat was in a flat-skull position in the stereotact. (B) Measurement of the outline of the cortex (*red curve*) in a sagittal Nissl-stained image from bat #3 (taken 1.6 mm lateral to midline), which was rotated to match the rotation of the brain of bat #2 in panel A. This *red outline* was then used to determine the vertical dorso-ventral (D-V) shift of each atlas coronal plate (see text for details). *Red arrow* denotes the highest D-V point of the brain: the top of the *skull* at this location was defined as the zero D-V coordinate of the atlas (see text; this is around atlas plates 44–49).

highest point of the *brain* in the atlas at 1.6-mm M-L has a D-V coordinate of −0.45 mm. Note that in this procedure we neglected the thickness of the meninges.

## Accuracy of the stereotaxic coordinates

We found that the shrinkage of the brain and of the sections was negligible—no more than a few percent. This was verified using three different approaches. First, we compared the A-P distance between the needle tracks, measured in two ways: (i) from the photograph of the whole brain taken immediately after the brain extraction (Fig. 2B) and (ii) stereotaxic distance between the pins as measured during surgery. This yielded an estimated shrinkage of < 5%. Second, we checked that there is no shrinkage of the individual sections, by comparing three measurements: (i) the M-L distance between the four calibration pins measured from the photograph of the whole brain taken immediately after the brain extraction (Fig. 2B); (ii) the M-L distance measured in vivo during surgery using the stereotact; and (iii) the M-L distance measured within the Nissl-stained mounted section, using the needle tracks that were readily identifiable in the photographs of the sections. The lack of shrinkage (estimated shrinkage < 5%) was true for the Nissl sections and AChE sections, but we noted that some of the Parvalbumin sections did exhibit shrinkage, and we therefore digitally expanded them in the atlas, to match them to the size of the adjacent Nissl sections. Third and finally, we performed many stereotaxic calibration surgeries, primarily in the dorsal hippocampus, based on the Nissl sections of the atlas, which allowed us to determine that the M-L coordinates (relative to midline) and A-P coordinates (relative to the sinus), as listed in the atlas, are correct for the dorsal hippocampus: the animal-to-animal errors were always less than 0.5 mm. However, when going more anteriorly than dorsal hippocampus, the A-P errors can be slightly larger than 0.5 mm, because we did not calibrate those regions systematically. In prefrontal cortex, for example, we found errors of up to 0.75 mm in A-P coordinates (where we tended to reach areas a few hundred μm more posterior than we aimed for). Therefore it is highly advised to perform separate calibrations of the A-P coordinates for each brain area of interest.

We do not expect substantial systematic stereotaxic differences based on bat weight because of the following considerations. Adult Egyptian fruit bats used for research are typically larger than average, and range in weight usually between 145 and 180 g. The bat used for this atlas (143.3 g) was at the lower end of this typical range of weights for experimental bats (although it was averaged sized for the species). However, we note that the atlas bat deviated in weight only by ~ 25% from the largest bats used for research. We did not study systematically the stereotaxic errors as a function of bat weight, but we expect the errors to be small, for the following reason. A weight deviation of 25% is quite a small deviation—much smaller, for example, than the ~ 80% weight deviation of large mature rats from the 290 g weight of the medium-sized rat that was used by Paxinos and Watson for their atlas (Paxinos and Watson, 2007). Paxinos and Watson showed that for a rat that is 50% higher in weight than their atlas rat, the A-P coordinates of relatively anterior brain regions did not deviate more than 0.7 mm. We therefore expect that for our much smaller potential weight differences, of up to 25%, these weight-related deviations will be below 0.5 mm.

Finally, we note that this atlas is based on a female brain, while most research is done on male Egyptian fruit bats. Importantly, all the injections and electrolytic lesions that we did over the last decade to verify the atlas coordinates were all carried out in male bats; therefore, we consider the atlas to be well representing the stereotaxic coordinates of the male brain. All appendixes in this atlas are based on male brains (except Appendix 1, which is from the atlas brain, and Appendix 7).

# Brain shape and brain weight

The brain of the Egyptian fruit bat is pear shaped, being wider at the back than in the front (Figs. 1 and 2A)—similar in shape to the guinea pig brain, and unlike the shape of the rat or mouse brain. The brain weight was not measured in the bat on which the atlas was based, but it was subsequently measured in seven male bats that were slightly larger (bat weights 161–170 g and one bat with a 186 g weight): the brain weight, after perfusion with 4% paraformaldehyde, was between 2.0 and 2.4 g, with an average of 2.15 g. This is similar to the weight of a male rat brain, which was reported to be 2.1 g on average after perfusion with 4% paraformaldehyde (Ray and Brecht, 2016), and ~ 2.0–2.5 g without perfusion (Donaldson and Hatai, 1932). We note that the brain weights that we measured for the bats may be biased upwards, because perfusion introduces extra fluids into the brain, which adds weight; therefore, the bat brain weight in vivo, without perfusion, is likely smaller than 2.0–2.4 g.

## Delineation of structures

We opted not to delineate the border of most structures because the borders are the most difficult to determine and require in many cases extensive tract-tracing data, which we did not have, except for a few specific structures such as the hippocampal formation and the auditory system. We therefore marked in most cases only the center of the structures, and have put borders only where we were very confident about them, which in part reflects our own scientific interests.

Nomenclature in this atlas follows generally the nomenclature of the Paxinos and Watson rat brain atlas (Paxinos and Watson, 2007). In particular, abbreviations of brain nuclei and cortical regions start with capital letters, while abbreviations of fiber bundles start with lower case letters; thus IC and ic should not be mixed up—the capitalized IC stands for inferior colliculus, while the lowercase ic stands for internal capsule. Notable exceptions from the Paxinos and Watson nomenclature include: (i) The retrosplenial and posterior cingulate cortex, where we in part followed the nomenclature of the primate brain, based on the detailed cytoarchitectonic description for the rhesus monkey (Kobayashi and Amaral, 2000), taking into account available descriptions of the cytoarchitectonics of the rodent retrosplenial cortex (Sugar et al., 2011) and cingulate cortex (Jones et al., 2005) (for further details see later). (ii) The anterior cingulate and medial prefrontal cortex were delineated based on cytoarchitectonic criteria, which nicely resemble those in rodents (see details later). The perirhinal cortex was subdivided into areas 35 and 36 (denoted PRh35 and PRh36), following our nomenclature as described previously in rodents (Kjonigsen et al., 2011; van Strien et al., 2009) (see details later). This differs from the definition given by (Paxinos and Watson, 2007), where area 35 is called perirhinal cortex and area 36 is referred to as ectorhinal cortex. We further differentiated the caudal portion of the perirhinal cortex, referring to it as postrhinal cortex (Burwell, 2001). (iii) The amygdala, where we refrained from detailed subdivisions, only separating between a lateral, basal, and central component, as well as accessory basal nucleus (see later). For ease of use, we listed the abbreviations alongside the full structure names on each plate, at the bottom of the page, as well as in two Indexes: Index of Structures, and Index of Abbreviations.

Following is a brief description of the rationale for the delineation of structures in this atlas.

### Basal ganglia and basal forebrain

We used TH staining (Appendix 2), ChAT staining (Appendix 3), and Parvalbumin staining (atlas plates) for identifying the caudate and putamen, globus pallidus, ventral pallidum, nucleus basalis, substantia nigra (SNC and SNR), ventral tegmental area (VTA), and medial forebrain bundle (mfb). These delineations were done on the following basis. For delineating the ventral pallidum and nucleus basalis, we used ChAT staining (Appendix 3) (Root et al., 2015) and also relied on a previous neuroanatomical study on the distribution of cholinergic neurons in the brain of this bat species (Maseko et al., 2007). For delineating the substantia nigra (SNC and SNR), ventral tegmental area (VTA), and medial forebrain bundle (mfb), we used the TH staining (see Appendix 2). Note that in this bat species, the VTA nuclei in the two hemispheres are merged together along the midline, spreading laterally, while in the rat and mouse the VTA of the two hemispheres are well separated into two independent nuclei. For delineating the caudate and putamen, we used AChE staining (see atlas plates) as well as TH staining (Appendix 2). Note that the ic (internal capsule) is separating the caudate and the putamen from each other: this is similar to how they are arranged in the primate, and very different from rodents, where the caudate and putamen are not separated. The globus pallidus was delineated using ChAT and TH staining (Appendix 3).

### Septal complex, and hypothalamus

The septal complex was subdivided using cytoarchitectonic criteria into the lateral and medial septum. We did not further subdivide the lateral septum, but differentiated the septohippocampal nucleus. Medial septum (MS), lateral septum (LS), and the diagonal band of Broca were identified based on ChAT staining (Appendix 3) and AChE staining (atlas plates), and subdivisions were named according to cytoarchitectonic criteria described in rodents (nuclei of the horizontal limb and vertical limb of the diagonal band—HDB and VDB) (Paxinos, 2014; Pessoa et al., 2019). We used oxytocin staining (Appendix 4) for identifying the paraventricular hypothalamic nucleus (PVH) and supraoptic nucleus (SO), both of which contain many oxytocin neurons.

## Amygdala

The amygdala is a complex structure that developmentally comprises cortical and subcortical elements. Whereas the subcortical elements are well preserved across species, the cortical ones show an increasing complexity when comparing primates to rodents (Olucha-Bordonau et al., 2014; Pessoa et al., 2019). Although different compartments of the amygdala are clearly differentially stained for AChE and Parvalbumin (see atlas plates), we confined ourselves to differentiating a lateral, basal, and central component (denoted as LA, BA, and CeA), as well as accessory basal nucleus (AB)—without providing further subdivisions, nor providing details on the parts of the extended amygdala and bed nucleus of the stria terminalis (Olucha-Bordonau et al., 2014, Pessoa et al., 2019).

## Thalamus

All thalamic nuclei were delineated using established cytoarchitectonic criteria for the rat (Vertes et al., 2014). To a large extent we follow the nomenclature for individual nuclei as given for the rat by (Paxinos and Watson, 2007). Regarding the posterior thalamus, we considered delineating a Pulvinar nucleus, located between the dLGN and LP thalamic nuclei, at the caudal portion of LP because an unidentified thalamic nucleus is clearly seen there, and it was previously identified as Pulvinar in a closely related bat species (Manger and Rosa, 2005). However, eventually we decided not to do so, and leave this nucleus unidentified, since we lack tract-tracing data as well as genetic profile data for this nucleus.

## Hippocampal region

The hippocampal region (Appendix 6) was considered to comprise the hippocampal formation and the parahippocampal region as defined previously for the rodent (Kjonigsen et al., 2011; van Strien et al., 2009). The hippocampal formation is further divided into the dentate gyrus, the three fields of the Cornu Ammonis (CA1, CA2, and CA3) and the subiculum. The parahippocampal region comprises the presubiculum, the parasubiculum, and the rhinal cortices, including the entorhinal, perirhinal, and postrhinal cortex. Further subdivisions are based on combined data, using Nissl stains and stains for PV immunohistochemistry as described previously for the rodent. The entorhinal cortex was divided into two, the lateral and medial entorhinal cortex. Although more elaborate divisions have been proposed, even in the case of bats (Gatome et al., 2010), we decided not to use them here, since functional studies in many species currently focus on differences between the two divisions (for a more detailed discussion, see (Nilssen et al., 2019)).

The delineations of the borders in the hippocampal region were based on cytoarchitectonics, as well as on extensive tracer injections into the various subfields of the hippocampal formation (Kleven et al., 2014). We verified the borders by checking self-consistency between the borders in the coronal sections (see atlas plates), sagittal sections (as in Appendix 7), and horizontal sections (as in Appendix 8), as well as 3D reconstructions of the hippocampal formation subregions (Appendix 6), which were based also on magnetic resonance imagining (MRI) scans of the bat brain (in another individual bat).

## Cerebral cortex

For the majority of cortical delineations, we used cytoarchitectonic and positional criteria, combined with chemoarchitecture, based on the distribution of PV (see atlas plates) and another protein, calretinin. Authors often have used the specific distribution of calbindin but in our hands calbindin immunohistochemistry did not produce reliable results in the Egyptian fruit bat, and calretinin worked much better. We have not tested other stains aside from the available AChE stain in the atlas plates. In a number of cases we obtained additional connectional data or other specific expression data: this will be described later for areas to which this applies.

*Primary auditory cortex* (Au1) was identified by injecting an anterograde tracer (biotinylated dextran amine, BDA) into the ventral part of the medial geniculate nucleus (MGV)—which is the lemniscal part of the auditory thalamus—and delineated the resulting staining in the cortex (see Appendix 5—first page, left column). *Ventral secondary auditory cortical area* (AuV) was identified by injecting BDA into the medial and dorsal parts of the medial geniculate nucleus (MGM and MGD; see Appendix 5—first page, right column). We verified these delineations by conducting double injections of BDA and cholera toxin (a retrograde tracer) into the cortex AuV (see Appendix 5—second page).

*Visual cortex* (VC) was identified based on a thick layer 4 using the Nissl staining, as well as by performing electrophysiological recordings of neuronal responses to visual stimuli (oriented bars), in experiments conducted by Michael Okun and Yonatan Katz in the lab of Ilan Lampl. The most V1-like neural responses (based on response latency, orientation tuning, and the occurrence of simple and complex cells) were at around 1.5 mm A-P (1.5 mm anterior to the transverse sinus) and 2–4 mm M-L.

*Motor cortex* (M) was identified based on a thick layer 5 using the Nissl staining.

*Somatosensory cortex* (SS) was identified based on a thick layer 4 in which Nissl staining showed a high packing density of relatively small-sized neuronal somata with a grain-like appearance.

*Posterior parietal cortex* (PPC) was delineated based on position, since it is well accepted that it is located between the dorsomedial part of the primary parietal (somatosensory) area and the occipital region. We used a combination of cyto- and chemoarchitectural criteria established in rodents—both rats and mice (Hovde et al., 2019; Olsen and Witter, 2016). In short, the rostrally adjacent primary somatosensory cortex and the caudally adjacent primary and secondary visual cortex both show a prominent layer 4 and a somewhat sparsely populated layer 5. In contrast, PPC shows a very poorly differentiable layer 4, and layer 5 is quite homogeneous and densely populated with pyramidal neurons. In many species, PPC has been subdivided into a more posterior area 7 and a more anteriorly located area 5, but as extensively discussed in Olsen and Witter (2016), area 5 is apparently not present in rodents. In the current version of the atlas, we have not further subdivided PPC; more connectional data would be needed to do so (see also the extensive discussion in Olsen and Witter, 2016).

*Retrosplenial cortex* was delineated based on the detailed cytoarchitectonic description for the rhesus monkey (Kobayashi and Amaral, 2000), taking into account available descriptions of the cytoarchitectonics of the rodent retrosplenial cortex (Sugar et al., 2011). We relied here on the monkey data because the retrosplenial cortex seems highly developed in the Egyptian fruit bat, resulting in more identifiable subdivisions than have been described in rodents. We differentiated between retrosplenial areas A29a/b and A29c, where A29a/b likely corresponds to A29l in the monkey and A29c corresponds to A29m in the monkey. We further differentiated A30 and A23. Whereas the architectonic criteria for A29 and its subdivisions and A30 are remarkably similar between rodents, the rhesus monkey, and the bat, rodents lack a clearly defined A23, but we feel it is recognizable in the bat. However, we did not further subdivide area 23, as has been done in the monkey. A29a/b forms the most ventral subdivision, and in rostral levels it covers a substantial part of the medial surface of the posterior part of the cingulate gyrus. A29c and A23 are located within the cingulate sulcus, whereas A29c mainly occupies the ventral bank moving into the fundus, and area 23 is located in the dorsal part of the fundus, extending along part of the dorsal bank. At more caudal levels, all cortical areas gradually shift to a more dorsal position such that area 23 is located on the medial surface of the brain just above the cingulate sulcus, area 30 covers the dorsal bank of that sulcus, and area 29c is part of the ventral bank. Cytoarchitectonically, the most ventrally positioned A29a differs from the dorsally adjacent A29b since it lacks a fully differentiated layered structure; A29a has a homogenous layer 2/3, while in A29b this layer is divided into a thin superficial, densely packed zone and a less dense deeper zone. The clearest differentiation between A29b and A29c is the presence of the dense granular-like layer 2 in A29b that is less apparent in A29c. A29c has no clear granular layer 4, in contrast to A30 and A23. The latter area shows a fully developed six-layered structure with a clearly visible layer 4 and subdivisions within layer 5.

*Perirhinal cortex and postrhinal cortex* were identified based on the descriptions in the rodent (Boccara et al., 2015; Burwell, 2001), in which we heavily rely on the distribution of Parvalbumin and calretinin in addition to cytoarchitectonic criteria. Note that in contrast to the nomenclature used by Paxinos and Watson, we elected to divide the perirhinal cortex along the rostro-caudal axis, introducing the usage of postrhinal cortex (Burwell, 2001). Both perirhinal and postrhinal cortex can be subdivided into a dorsal and a ventral subdivision, as described in detail by (Burwell, 2001). For the ventral part of the perirhinal cortex, we used PRh35 (area 35), in contrast to the dorsal part, which is denoted as PRh36 (area 36; note that according to the nomenclature of Paxinos and Watson, these areas are called perirhinal and ectorhinal cortex, respectively). The postrhinal cortex also has this division into a ventral and dorsal subdivisions (denoted here PORd and PORv).

*Prefrontal areas*—in particular cingulate cortex (Cg), prelimbic cortex (PrL), and infralimbic cortex (IL)—were delineated as follows (courtesy of

Saikat Ray). The dorsal and ventral borders of these medial prefrontal regions were identified based on cytoarchitecture, histochemistry, immunohistochemistry, and tract tracing. The most ventral area is the infralimbic cortex, bordered more dorsally by the prelimbic cortex, followed by the most anterior part of the cingulate cortex or anterior area 24 (Jones et al., 2005; Van De Werd et al., 2010). The borders of the *prelimbic cortex* were delineated based on differences in cytoarchitecture, determined by Nissl staining. Similar to what has been described in rodents, it has a sparser cell density in layer 3 than the regions around it—the cingulate cortex, which lies dorsal to it, and the infralimbic cortex, lying ventral to it (Jones et al., 2005; Van De Werd et al., 2010; Vogt, 2014). AChE histochemistry also confirms these borders, as the prelimbic cortex has high cholinergic staining in layer 2, but not in layers 1 and 3; by contrast, both the flanking regions of cingulate cortex and infralimbic cortex have a uniform cholinergic staining across all layers. The borders of the *infralimbic cortex* were delineated based on immunohistochemical stainings for the transmembrane protein Wolframin (Wolfram Syndrome 1, WFS1; 1:200; Proteintech), the proteolipid Neuronatin (NNat; 1:200; Abcam), and the calcium binding protein Parvalbumin (PV). The infralimbic cortex has an enriched expression of WFS1 and NNat across its layers, unlike the prelimbic cortex dorsal to it and the dorsal peduncular cortex ventral to it. Additionally, the infralimbic cortex has a lower PV expression than the surrounding regions. The location of the infralimbic cortex was also confirmed by tract-tracing experiments in three bats, by injecting the anterograde tracer dextran amine (10k) in the ventral hippocampus, which in rodents projects specifically to the infralimbic cortex but not to the surrounding areas (Jay and Witter, 1991; Swanson, 1981).

The *insular cortex* is a complex area, both functionally as well as in terms of overall connectivity and cytoarchitectonic features. It extends from the orbitofrontal cortex rostrally to the perirhinal cortex caudally, overlying the claustrum (Cl). The insular cortex is generally considered to be divided into anterior, posterior, and parietal insular domains. In each of these insular domains, granular, dysgranular, and agranular insular divisions are present (Mathiasen et al., 2015). In view of the immense structural and connectional diversity of the insular cortex (Butti and Hof, 2010), and since we lack connectional data, we have only indicated the approximate position of the insular cortex (In) in the atlas.

The *piriform cortex* (Pir), or primary olfactory cortex in the mammalian brain, is typically described as a three-layered cortex comprising a single pyramidal cell layer, populated with very large pyramidal neurons, covered with a well-developed molecular layer. Deep to the pyramidal cell layer one finds a polymorph layer composed of a diverse population of neurons. Deep to this the endopiriform nucleus (En) is found. A striking feature of the piriform cortex is the evident distribution of the olfactory tract in the superficial part of the molecular layer. This architectonic feature is so evident in any mammalian brain that we have systematically indicated the borders of the piriform cortex, although without further subdividing it.

# References


Boccara CN, Kjonigsen LJ, Hammer IM, Bjaalie JG, Leergaard TB, Witter MP. A three-plane architectonic atlas of the rat hippocampal region. Hippocampus 2015;25:838–57.

Burwell RD. Borders and cytoarchitecture of the perirhinal and postrhinal cortices in the rat. J Comp Neurol 2001;437:17–41.

Butti C, Hof PR. The insular cortex: a comparative perspective. Brain Struct Funct 2010;214:477–93.

Donaldson HH, Hatai S. On the weight of the parts of the brain and on the percentage of water in them according to brain weight and to age, in albino and wild Norway rats. J Comp Neurol 1932;53:263–307.

Gatome CW, Slomianka L, Mwangi DK, Lipp HP, Amrein I. The entorhinal cortex of the Megachiroptera: a comparative study of Wahlberg's epauletted fruit bat and the straw-coloured fruit bat. Brain Struct Funct 2010;214:375–93.

Hovde K, Gianatti M, Witter MP, Whitlock JR. Architecture and organization of mouse posterior parietal cortex relative to extrastriate areas. Eur J Neurosci 2019;49:1313–29.

Jay TM, Witter MP. Distribution of hippocampal CA1 and subicular efferents in the prefrontal cortex of the rat studied by means of anterograde transport of *Phaseolus vulgaris*-leucoagglutinin. J Comp Neurol 1991;313:574–86.

Jones BF, Groenewegen HJ, Witter MP. Intrinsic connections of the cingulate cortex in the rat suggest the existence of multiple functionally segregated networks. Neuroscience 2005;133:193–207.

Kjonigsen LJ, Leergaard TB, Witter MP, Bjaalie JG. Digital atlas of anatomical subdivisions and boundaries of the rat hippocampal region. Front Neuroinform 2011;5:2.

Kleven H, Gatome CW, Las L, Ulanovsky N, Witter MP. FENS Forum; 2014.

Kobayashi Y, Amaral DG. Macaque monkey retrosplenial cortex: I. Three-dimensional and cytoarchitectonic organization. J Comp Neurol 2000;426:339–65.

Manger PR, Rosa MG. Visual thalamocortical projections in the flying fox: parallel pathways to striate and extrastriate areas. Neuroscience 2005;130:497–511.

Maseko BC, Bourne JA, Manger PR. Distribution and morphology of cholinergic, putative catecholaminergic and serotonergic neurons in the brain of the Egyptian rousette flying fox, *Rousettus aegyptiacus*. J Chem Neuroanat 2007;34:108–27.

Mathiasen ML, Hansen L, Witter MP. Insular projections to the parahippocampal region in the rat. J Comp Neurol 2015;523:1379–98.

Nilssen ES, Doan TP, Nigro MJ, Ohara S, Witter MP. Neurons and networks in the entorhinal cortex: a reappraisal of the lateral and medial entorhinal subdivisions mediating parallel cortical pathways. Hippocampus 2019;29:1238–54.

Olsen GM, Witter MP. Posterior parietal cortex of the rat: architectural delineation and thalamic differentiation. J Comp Neurol 2016;524:3774–809.

Olucha-Bordonau FE, Fortes-Marco L, Otero-García M, Lanuza E, Martínez-García. Amygdala: structure and function. In: Paxinos G, editor. The rat nervous system. 4th ed. Academic Press; 2014.

Paxinos G. The rat nervous system. 4th ed. Academic Press; 2014.

Paxinos G, Watson C. The rat brain in stereotaxic coordinates. 6th ed. Academic Press; 2007.

Pessoa L, Medina L, Hof PR, Desfilis E. Neural architecture of the vertebrate brain: implications for the interaction between emotion and cognition. Neurosci Biobehav Rev 2019;107:296–312.

Ray S, Brecht M. Structural development and dorsoventral maturation of the medial entorhinal cortex. eLife 2016;5, e13343.

Root DH, Melendez RI, Zaborszky L, Napier TC. The ventral pallidum: subregion-specific functional anatomy and roles in motivated behaviors. Prog Neurobiol 2015;130:29–70.

Schneider R. Gehirn von *Rousettus aegyptiacus*, E. Geoffroy 1810 (Megachiroptera, Chiroptera, Mammalia). Kramer; 1966.

Sugar J, Witter MP, van Strien NM, Cappaert NL. The retrosplenial cortex: intrinsic connectivity and connections with the (para)hippocampal region in the rat. An interactive connectome. Front Neuroinform 2011;5:7.

Swanson LW. A direct projection from Ammon's horn to prefrontal cortex in the rat. Brain Res 1981;217:150–4.

Van De Werd HJ, Rajkowska G, Evers P, Uylings HB. Cytoarchitectonic and chemoarchitectonic characterization of the prefrontal cortical areas in the mouse. Brain Struct Funct 2010;214:339–53.

van Strien NM, Cappaert NL, Witter MP. The anatomy of memory: an interactive overview of the parahippocampal-hippocampal network. Nat Rev Neurosci 2009;10:272–82.

Vertes RP, Linley SB, Groenewegen HJ, Witter MP. Thalamus. In: Paxinos G, editor. The rat nervous system. 4th ed. Academic Press; 2014.

Vogt BA. Cingulate cortex and pain architecture. In: Paxinos G, editor. The rat nervous system. 4th ed. Academic Press; 2014.

# Index of structures

Names of structures are listed in alphabetical order.
Each name is followed by the abbreviation of the structure.

|  |  |
|---|---|
| 3rd ventricle | 3V |
| Accessory basal nucleus amygdala | AB |
| Accumbens nucleus | Acb |
| Accumbens nucleus, core | AcbC |
| Accumbens nucleus, shell | AcbSh |
| Alveus of the hippocampus | alv |
| Amygdalo-entorhinal transition area | AE |
| Anterior commissure | ac |
| Anterior complex | AC |
| Anterior cortical amygdaloid area | ACo |
| Anterior hypothal area, ant part | AHA |
| Anterior olfactory area | AO |
| Anterior olfactory nucleus | AON |
| Anterior olfactory nucleus, dorsal part | AOD |
| Anterior olfactory nucleus, lateral part | AOL |
| Anterior olfactory nu, medial part | AOM |
| Anterior olfactory nu, ventral part | AOV |
| Anterior pretectal nucleus | APT |
| Anterior tegmental nucleus | ATg |
| Anterodorsal thalamic nucleus | AD |
| Aqueduct | Aq |
| Arcuate hypothalamic nucleus | Arc |
| Basal nucleus (nucleus basalis) | B |
| Bed nu stria terminalis | BST |
| Brachium of the inferior colliculus | bic |
| Caudal linear nucleus of the raphe | CLi |
| Caudate | C |
| Caudate putamen (striatum) | CPu |
| Central amygdaloid nucleus | CeA |
| Central medial thalamic nucleus | CM |
| Centrolateral thalamic nucleus | CL |
| Cerebral peduncle | cp |
| Cingulate cortex | Cg |
| Cingulum | cg |
| Claustrum | Cl |
| Commissure of the superior colliculus | csc |
| Corpus callosum | cc |
| Decuss sup cereb peduncle | xscp |
| Deep cerebral white matter | dcw |
| Deep mesencephalic nucleus | DpMe |
| Dentate gyrus | DG |
| Dorsal 3rd ventricle | D3V |

|  |  |
|---|---|
| Dorsal lateral geniculate nucleus | DLG |
| Dorsal nu of lateral lemniscus | DLL |
| Dorsal peduncular cortex | DP |
| Dorsal raphe nucleus | DR |
| Dorsomedial hypothalamic nucleus | DMH |
| Dorsomedial periaqueductal gray | DMPAG |
| Edinger-Westphal nucleus | EW |
| Endopiriform nucleus | En |
| Entorhinal cx-hippocampal transition | EC/Hi |
| Ependymal and subendymal layer | E |
| External capsule | ec |
| External plexiform layer of the olf bulb | EPl |
| Fasciculus retroflexus | fr |
| Fasciola cinereum | FC |
| Field CA1 of the hippocampus | CA1 |
| Field CA2 of the hippocampus | CA2 |
| Field CA3 of the hippocampus | CA3 |
| Fimbria of the hippocampus | fi |
| Forceps minor of the corpus callosum | fmi |
| Fornix | f |
| Frontal cortex | Fr |
| Globus pallidus | GP |
| Glomerular layer of the olfactory bulb | Gl |
| Granule cell layer of the olfactory bulb | GrO |
| Habenula | Hb |
| Horizontal limb nu diagonal band | HDB |
| Indusium griseum | IG |
| Inferior colliculus | IC |

|  |  |
|---|---|
| Inferior colliculus, central nucleus | ICC |
| Inferior colliculus, external nucleus | ICX |
| Infralimbic cortex | IL |
| Insular cortex | In |
| Interanteromedial thalamic nucleus | IAM |
| Intermediate nu of lateral lemniscus | ILL |
| Intermediodorsal thalamic nucleus | IMD |
| Internal capsule | ic |
| Internal plexiform layer of the olf bulb | IPl |
| Interpeduncular nucleus, apical | IPA |
| Interpeduncular nucleus, caudal | IPC |
| Interpeduncular nucleus, lateral | IPL |
| Interpeduncular nucleus, rostral | IPR |
| Island of Calleja, major island | ICjM |
| Lateral amygdaloid nucleus | LA |
| Lateral entorhinal cortex | LEC |
| Lateral hypothalamic area | LH |
| Lateral mammillary nucleus | LM |
| Lateral olfactory tract | lo |
| Lateral orbital cortex | LO |
| Lateral posterior thalamic nucleus | LP |
| Lateral septal nucleus | LS |
| Lateral ventricle | LV |
| Laterodorsal thalamic nucleus | LD |
| Mammillary peduncle | mp |
| Mammillothalamic tract | mt |
| Med amy nucleus, post vent part | MePV |
| Med geniculate nucleus | MG |

| Structure | Abbreviation |
|---|---|
| Med geniculate nucleus, dorsal part | MGD |
| Med geniculate nucleus, medial part | MGM |
| Med geniculate nucleus, ventral part | MGV |
| Med mamm nu, lat part | ML |
| Med mamm nu, med part | MM |
| Medial amygdaloid nucleus, ant part | MeA |
| Medial entorhinal cortex | MEC |
| Medial forebrain bundle | mfb |
| Medial lemniscus | ml |
| Medial longitudinal fasciculus | mlf |
| Medial orbital cortex | MO |
| Medial preoptic nucleus | MPO |
| Medial pretectal nucleus | MPT |
| Medial septal nucleus | MS |
| Median eminence | ME |
| Median preoptic nucleus | MnPO |
| Median raphe nucleus | MnR |
| Mediodorsal thalamic nucleus | MD |
| Middle cerebellar peduncle | mcp |
| Mitral cell layer of the olfactory bulb | Mi |
| Motor cortex | M |
| Nu stria terminalis | NST |
| Oculomotor nucleus | 3N |
| Olfactory tubercle | Tu |
| Olfactory ventricle | OV |
| Optic tract | opt |
| Parafascicular thalamic nucleus | PF |
| Parasubiculum | PaS |

| Structure | Abbreviation |
|---|---|
| Paratenial thalamic nucleus | PT |
| Paratrochlear nucleus | Pa4 |
| Paraventricular hypothalamic nucleus | PVH |
| Paraventricular thalamic nucleus | PV |
| Periaqueductal gray | PAG |
| Perirhinal cortex, area 35 | PRh35 |
| Perirhinal cortex, area 36 | PRh36 |
| Pineal recess | PiRe |
| Piriform cortex | Pir |
| Pontine nuclei | Pn |
| Pontine reticular nucleus, oral part | PnO |
| Posterior commissure | pc |
| Posterior hypothalamic area | PHA |
| Posterior hypothalamic nucleus | PH |
| Posterior parietal cortex | PPC |
| Posterior thalamic nu group, triangular | PoT |
| Posterior thalamic nuclear group | Po |
| Postmed cortical amy | PMCo |
| Postrhinal cortex | POR |
| Postrhinal cortex, dorsal | PORd |
| Postrhinal cortex, ventral | PORv |
| Prelimbic cortex | PrL |
| Presubiculum | PrS |
| Primary auditory cortex | Au1 |
| Putamen | Pu |
| Red nucleus | R |
| Reticular thalamic nucleus | Rt |
| Retrosplenial cortex | RS |

|  |  |
|---|---|
| Retrosplenial ctx area 23 | RS23 |
| Retrosplenial ctx area 29 | RS29 |
| Retrosplenial ctx area 29ab | RS29ab |
| Retrosplenial ctx area 29c | RS29c |
| Retrosplenial ctx area 30 | RS30 |
| Reuniens thalamic nucleus | Re |
| Rhinal fissure | rf |
| Rhomboid thalamic nucleus | Rh |
| Rostral linear nucleus (midbrain) | RLi |
| Secondary auditory ctx ventral area | AuV |
| Septofimbrial nucleus | SFi |
| Septohippocampal nucleus | SHi |
| Somatosensory cortex | SS |
| Stria medullaris | sm |
| Stria terminalis | st |
| Subfornical organ | SFO |
| Subgeniculate nu of thal | SubG |
| Subiculum | S |
| Substantia nigra, compact part | SNC |
| Substantia nigra, reticular part | SNR |
| Subthalamic nucleus | STh |
| Superior colliculus | SC |
| Suprachiasmatic nucleus | SCh |
| Supramammillary nu, medial part | SuMM |
| Supraoptic nucleus | SO |
| Tenia tecta | TT |
| Trochlear nucleus | 4N |
| Vent post lat thal nucleus | VPL |
| Vent post med thal nucleus | VPM |
| Ventral anterior thalamic nucleus | VA |
| Ventral hippocampal commissure | vhc |
| Ventral lateral thalamic nucleus | VL |
| Ventral nu of lateral lemniscus | VLL |
| Ventral orbital cortex | VO |
| Ventral pallidum | VP |
| Ventral tegmental area | VTA |
| Ventrolateral geniculate nucleus | VLG |
| Ventromedial hypothalamic nucleus | VMH |
| Ventromedial thalamic nucleus | VM |
| Vertical limb nu diagonal band | VDB |
| Visual cortex | VC |
| Zona incerta | ZI |

# Index of abbreviations

Abbreviations are listed in alphabetical order.
Each abbreviation is followed by the structure name, and by the plate numbers on which this structure appears.

# Plates

## Plate 1 - Sinus: 13.58 mm

Interaural: 16.28 mm
Nissl section 14

Interaural: 16.32 mm
AChE section 13

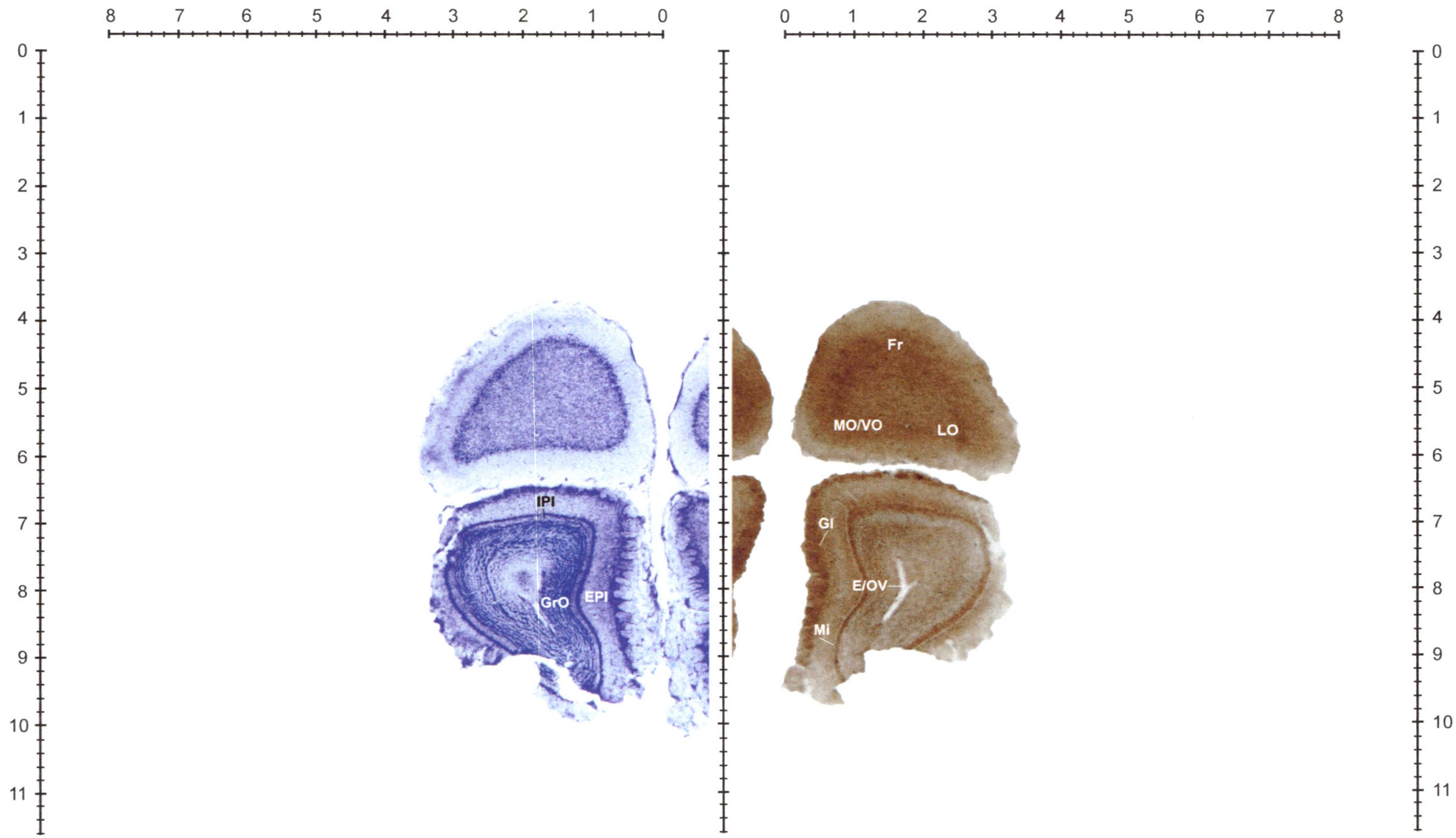


**E** ependymal and subendymal layer
**EPl** external plexiform layer of the olf bulb
**Fr** frontal cortex
**Gl** glomerular layer of the olfactory bulb
**GrO** granule cell layer of the olfactory bulb
**IPl** internal plexiform layer of the olf bulb
**LO** lateral orbital cortex
**Mi** mitral cell layer of the olfactory bulb
**MO** medial orbital cortex
**OV** olfactory ventricle
**VO** ventral orbital cortex

## Plate 2 - Sinus: 13.42 mm

Interaural: 16.12 mm
Nissl section 16

Interaural: 16.07 mm
AChE section 16

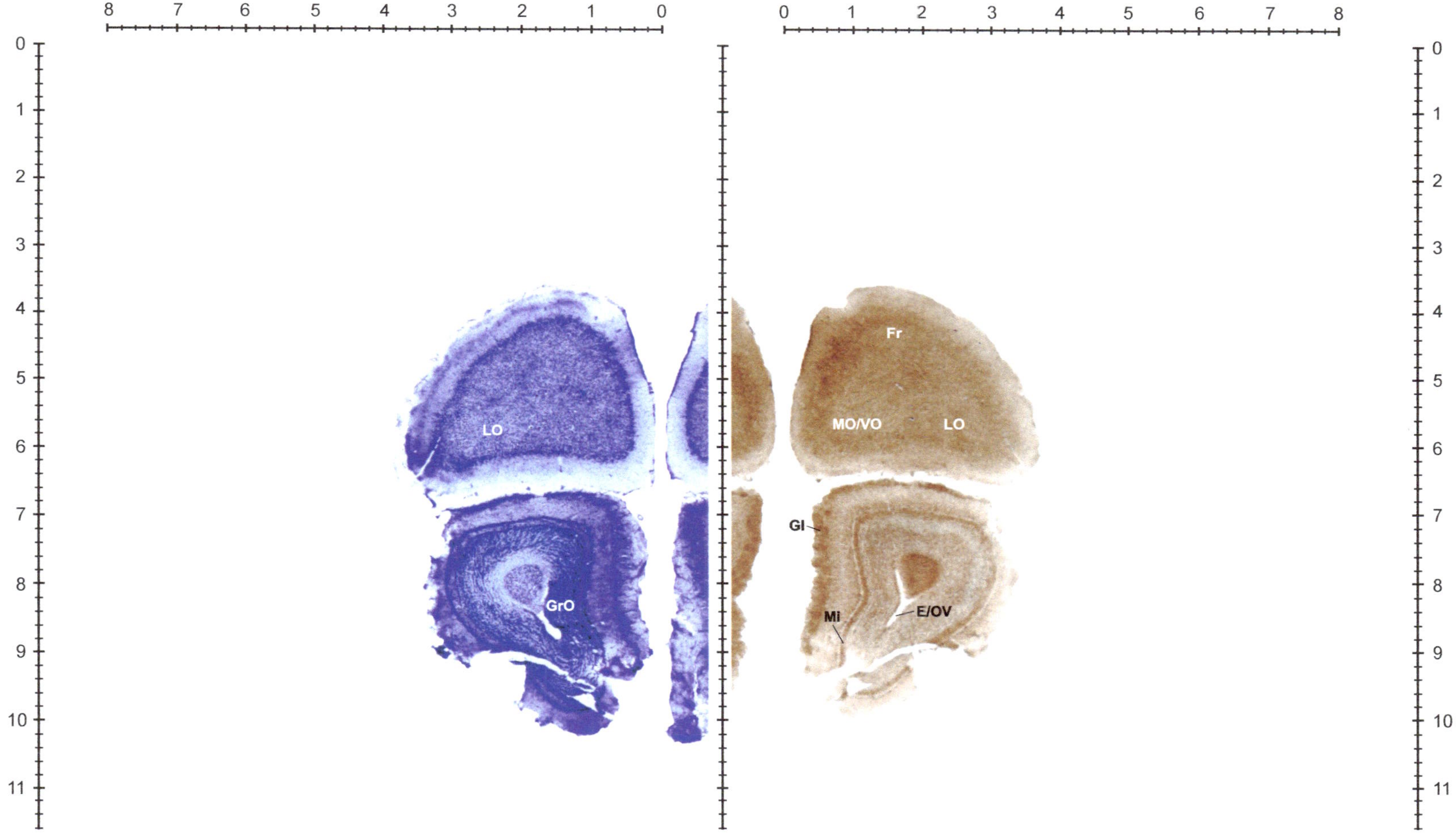


**E** ependymal and subendymal layer
**Fr** frontal cortex
**Gl** glomerular layer of the olfactory bulb
**GrO** granule cell layer of the olfactory bulb
**LO** lateral orbital cortex
**Mi** mitral cell layer of the olfactory bulb
**MO** medial orbital cortex
**OV** olfactory ventricle
**VO** ventral orbital cortex

## Plate 3 - Sinus: 13.33 mm

Interaural: 16.03 mm
Nissl section 17

Interaural: 15.99 mm
Parv section 17

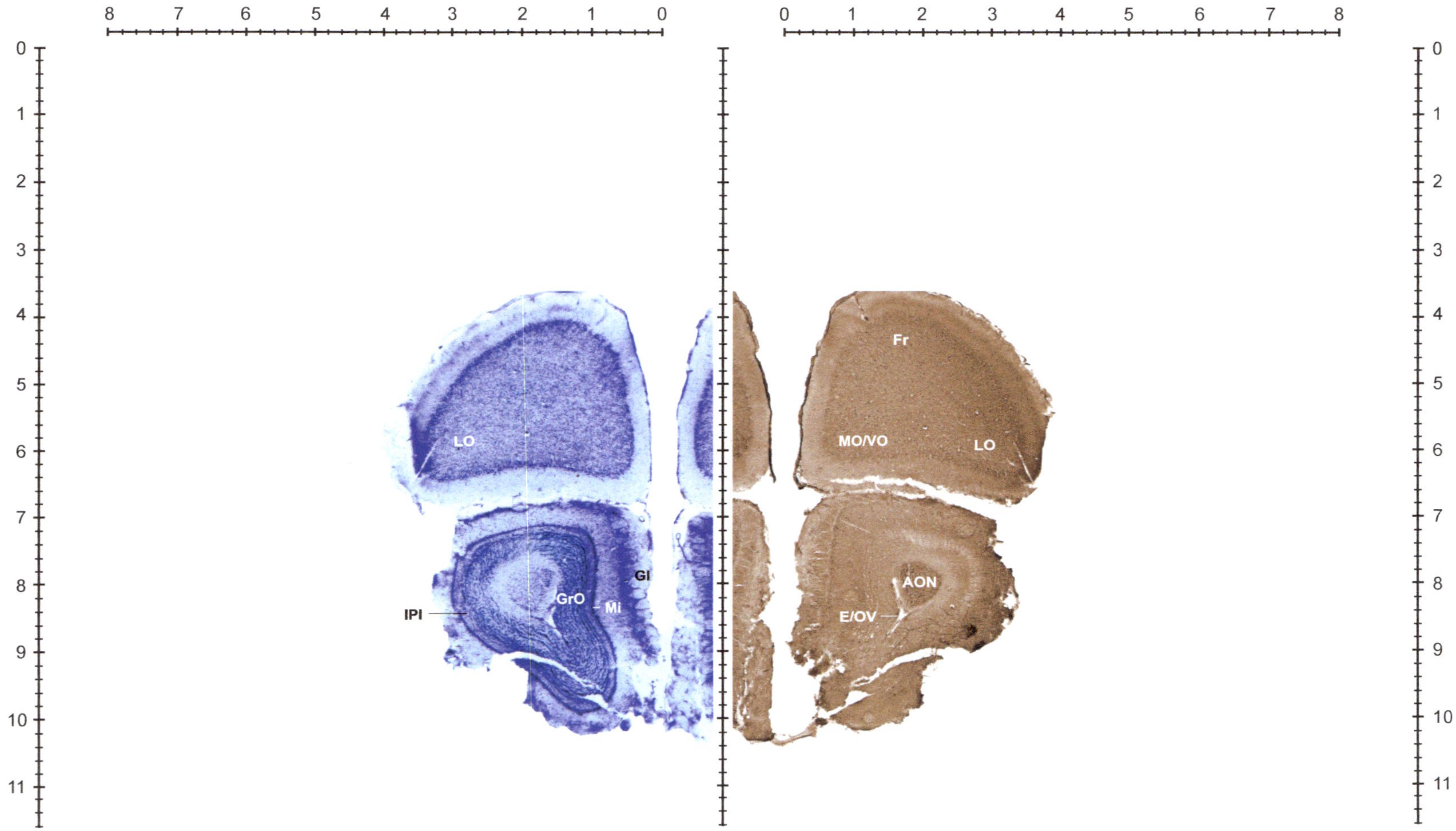


**AON** anterior olfactory nucleus
**E** ependymal and subendymal layer
**Fr** frontal cortex
**Gl** glomerular layer of the olfactory bulb
**GrO** granule cell layer of the olfactory bulb
**IPl** internal plexiform layer of the olf bulb
**LO** lateral orbital cortex
**Mi** mitral cell layer of the olfactory bulb
**MO** medial orbital cortex
**OV** olfactory ventricle
**VO** ventral orbital cortex

**Plate 4 - Sinus: 13.09 mm**

Interaural: 15.79 mm

Nissl section 20

Interaural: 15.83 mm

AChE section 19

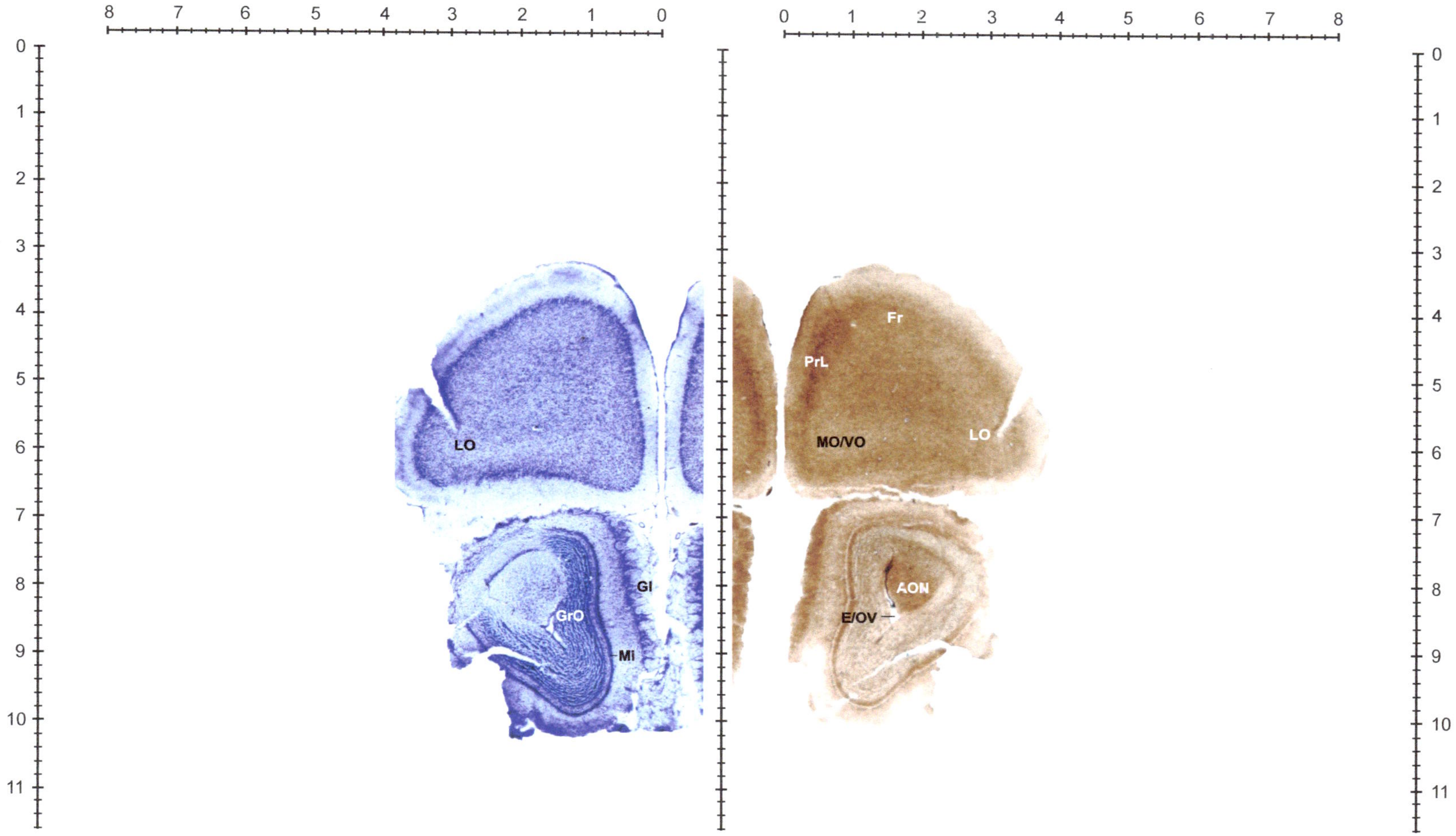


**AON** anterior olfactory nucleus
**E** ependymal and subendymal layer
**Fr** frontal cortex
**Gl** glomerular layer of the olfactory bulb
**GrO** granule cell layer of the olfactory bulb
**LO** lateral orbital cortex
**Mi** mitral cell layer of the olfactory bulb
**MO** medial orbital cortex
**OV** olfactory ventricle
**PrL** prelimbic cortex
**VO** ventral orbital cortex

## Plate 5 - Sinus: 12.92 mm

Interaural: 15.62 mm
Nissl section 22

Interaural: 15.66 mm
AChE section 21

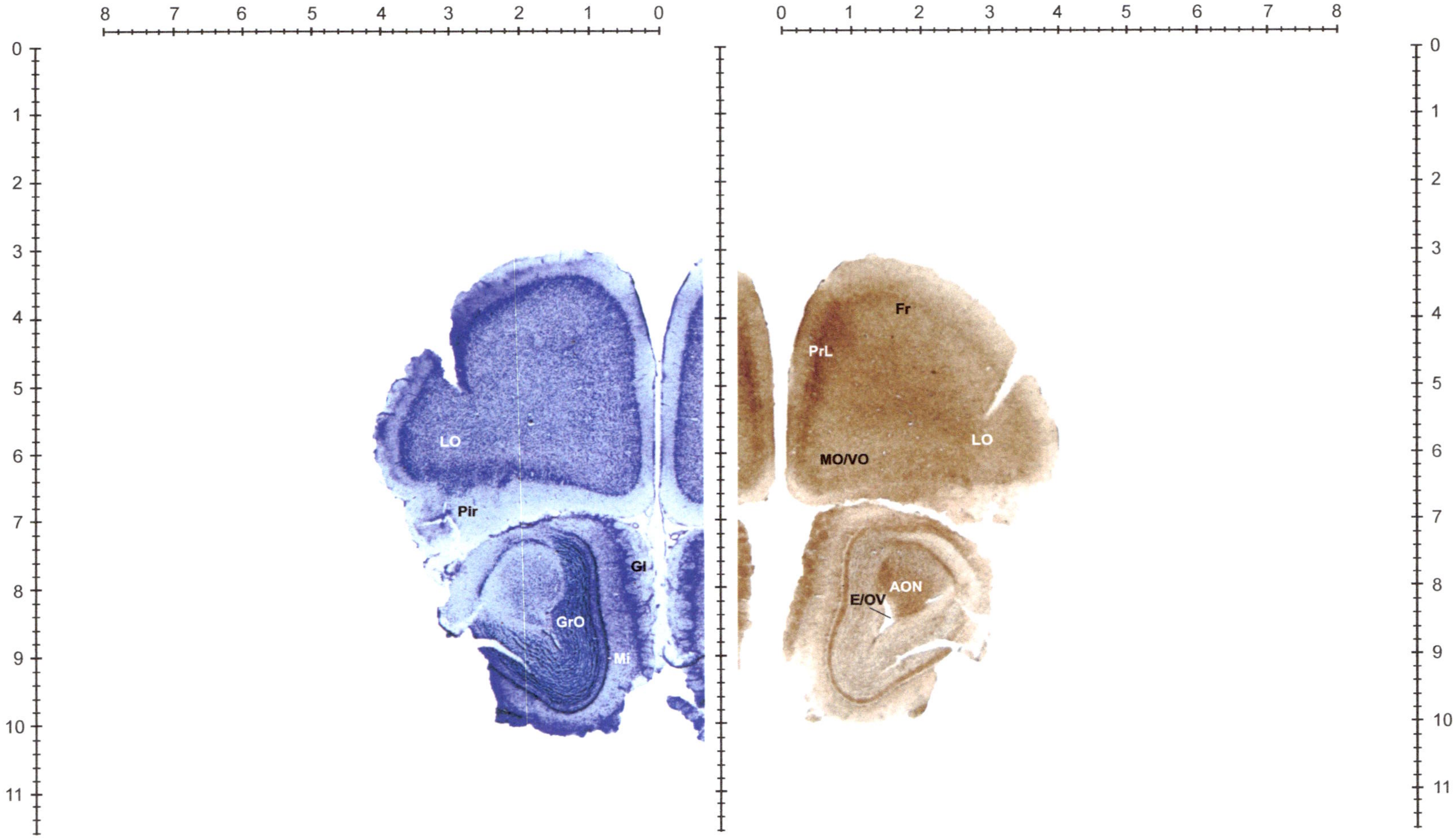


**AON** anterior olfactory nucleus
**E** ependymal and subendymal layer
**Fr** frontal cortex
**Gl** glomerular layer of the olfactory bulb
**GrO** granule cell layer of the olfactory bulb
**LO** lateral orbital cortex
**Mi** mitral cell layer of the olfactory bulb
**MO** medial orbital cortex
**OV** olfactory ventricle
**PrL** prelimbic cortex
**Pir** piriform cortex
**VO** ventral orbital cortex

## Plate 6 - Sinus: 12.84 mm

Interaural: 15.54 mm
Nissl section 23

Interaural: 15.42 mm
AChE section 24

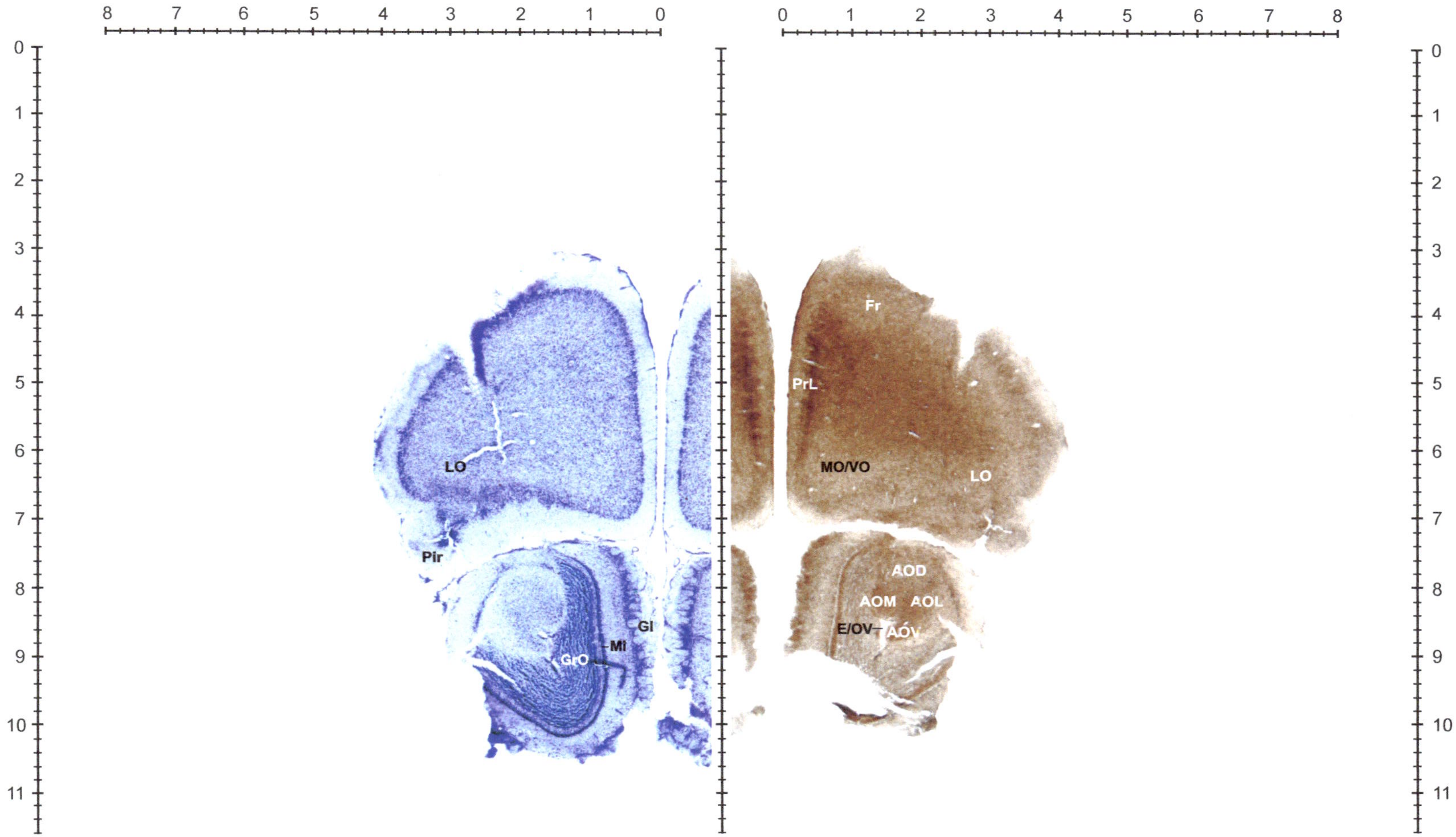


**AOD** anterior olfactory nucleus, dorsal part
**AOL** anterior olfactory nucleus, lateral part
**AOM** anterior olfactory nu, medial part
**AOV** anterior olfactory nu, ventral part
**E** ependymal and subendymal layer
**Fr** frontal cortex
**Gl** glomerular layer of the olfactory bulb
**GrO** granule cell layer of the olfactory bulb
**LO** lateral orbital cortex
**Mi** mitral cell layer of the olfactory bulb
**MO** medial orbital cortex
**OV** olfactory ventricle
**Pir** piriform cortex
**PrL** prelimbic cortex
**VO** ventral orbital cortex

## Plate 7 - Sinus: 12.60 mm

Interaural: 15.30 mm
Nissl section 26

Interaural: 15.25 mm
Parv section 26

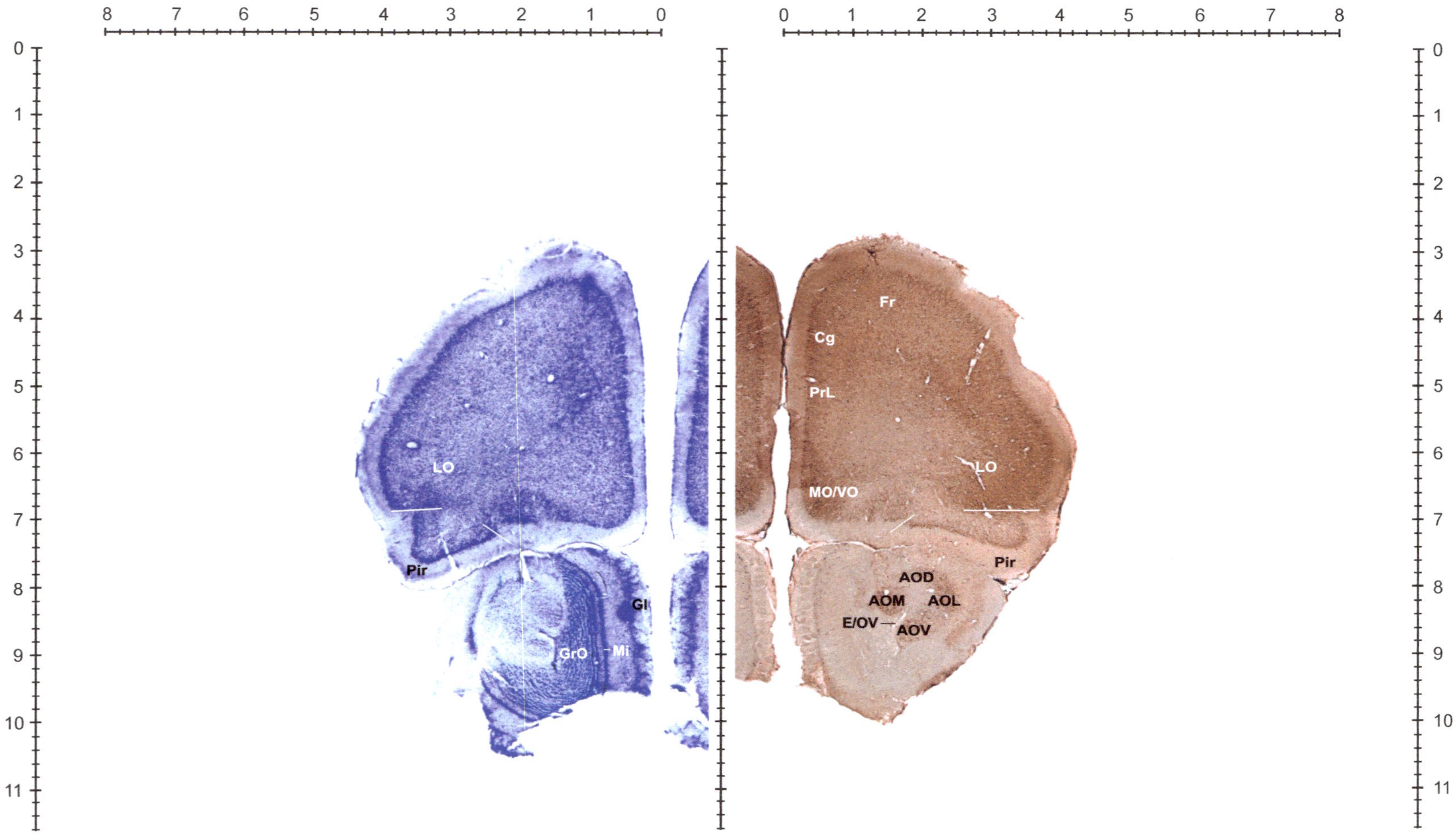


**AOD** anterior olfactory nucleus, dorsal part
**AOL** anterior olfactory nucleus, lateral part
**AOM** anterior olfactory nu, medial part
**AOV** anterior olfactory nu, ventral part
**Cg** cingulate cortex
**E** ependymal and subendymal layer
**Fr** frontal cortex
**Gl** glomerular layer of the olfactory bulb
**GrO** granule cell layer of the olfactory bulb
**LO** lateral orbital cortex
**Mi** mitral cell layer of the olfactory bulb
**MO** medial orbital cortex
**OV** olfactory ventricle
**Pir** piriform cortex
**PrL** prelimbic cortex
**VO** ventral orbital cortex

**Plate 8 - Sinus: 12.43 mm**

Interaural: 15.13 mm
Nissl section 28

Interaural: 15.09 mm
AChE section 28

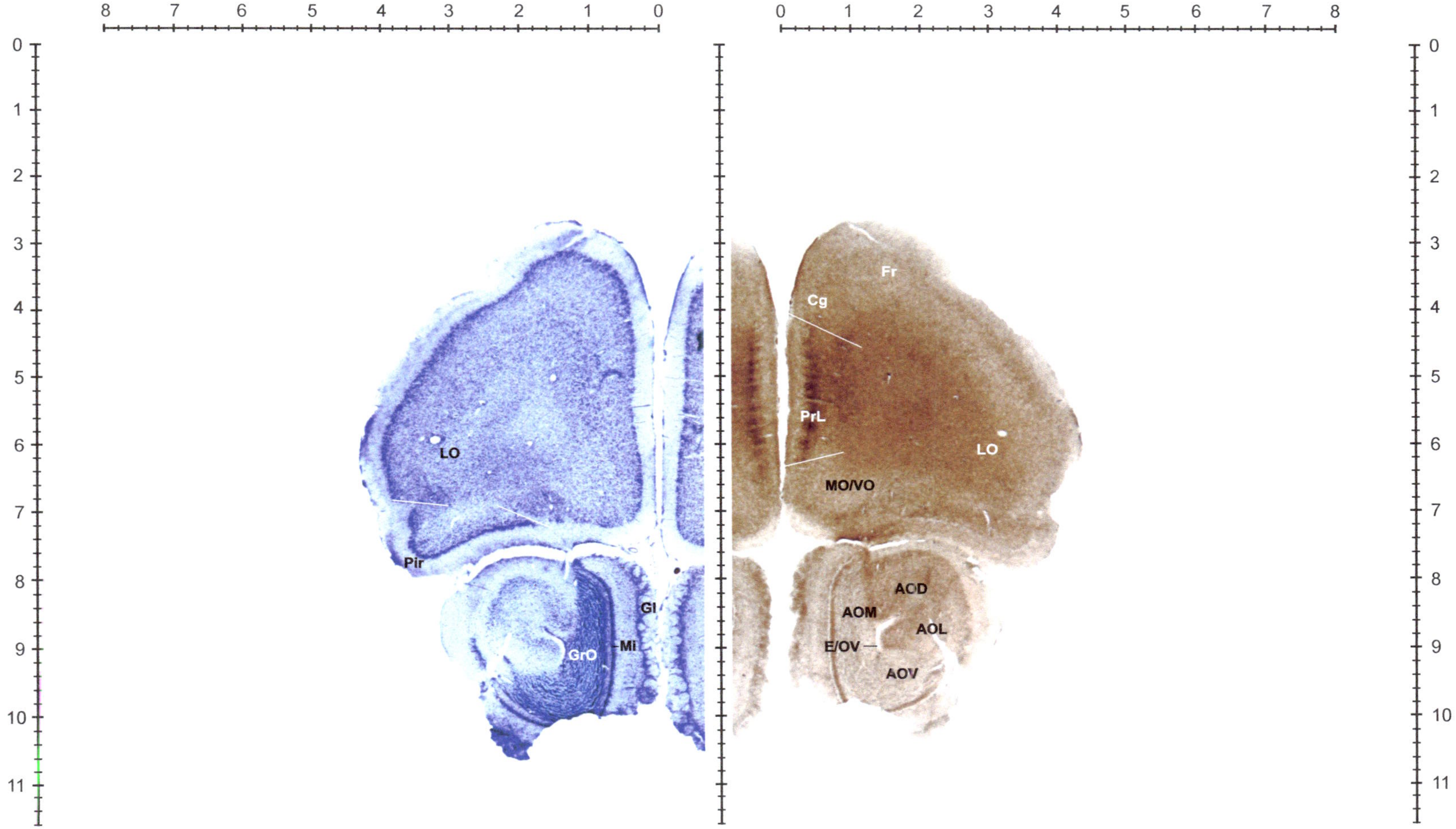


**AOD** anterior olfactory nucleus, dorsal part
**AOL** anterior olfactory nucleus, lateral part
**AOM** anterior olfactory nu, medial part
**AOV** anterior olfactory nu, ventral part
**Cg** cingulate cortex
**E** ependymal and subendymal layer
**Fr** frontal cortex
**Gl** glomerular layer of the olfactory bulb
**GrO** granule cell layer of the olfactory bulb
**LO** lateral orbital cortex
**Mi** mitral cell layer of the olfactory bulb
**MO** medial orbital cortex
**OV** olfactory ventricle
**Pir** piriform cortex
**PrL** prelimbic cortex
**VO** ventral orbital cortex

## Plate 9 - Sinus: 12.27 mm

Interaural: 14.97 mm
Nissl section 30

Interaural: 14.93 mm
AChE section 30

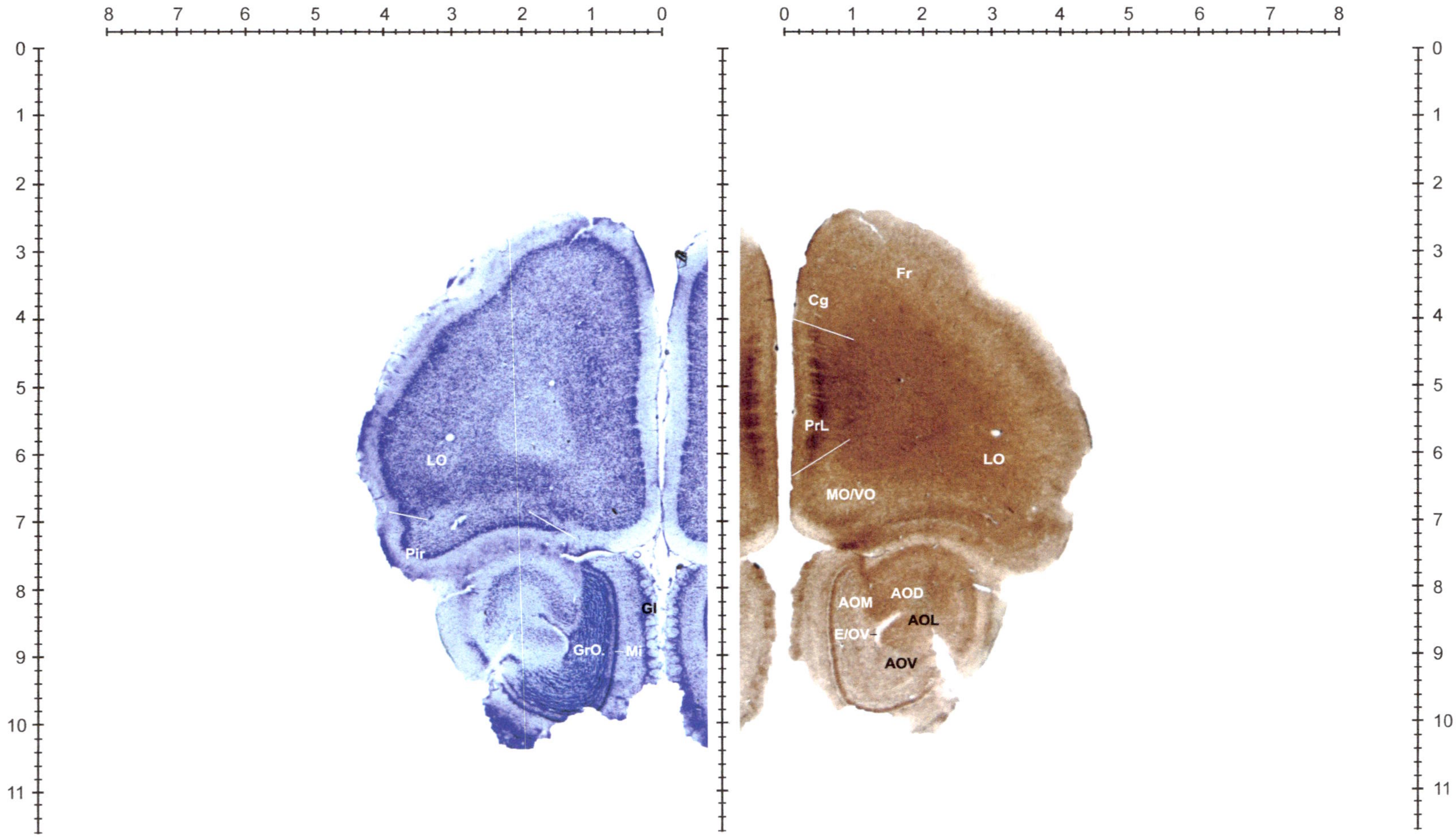


**AOD** anterior olfactory nucleus, dorsal part
**AOL** anterior olfactory nucleus, lateral part
**AOM** anterior olfactory nu, medial part
**AOV** anterior olfactory nu, ventral part
**Cg** cingulate cortex
**E** ependymal and subendymal layer
**Fr** frontal cortex
**Gl** glomerular layer of the olfactory bulb
**GrO** granule cell layer of the olfactory bulb
**LO** lateral orbital cortex
**Mi** mitral cell layer of the olfactory bulb
**MO** medial orbital cortex
**OV** olfactory ventricle
**Pir** piriform cortex
**PrL** prelimbic cortex
**VO** ventral orbital cortex

**Plate 10 - Sinus: 12.10 mm**

Interaural: 14.80 mm
Nissl section 32

Interaural: 14.76 mm
Parv section 32

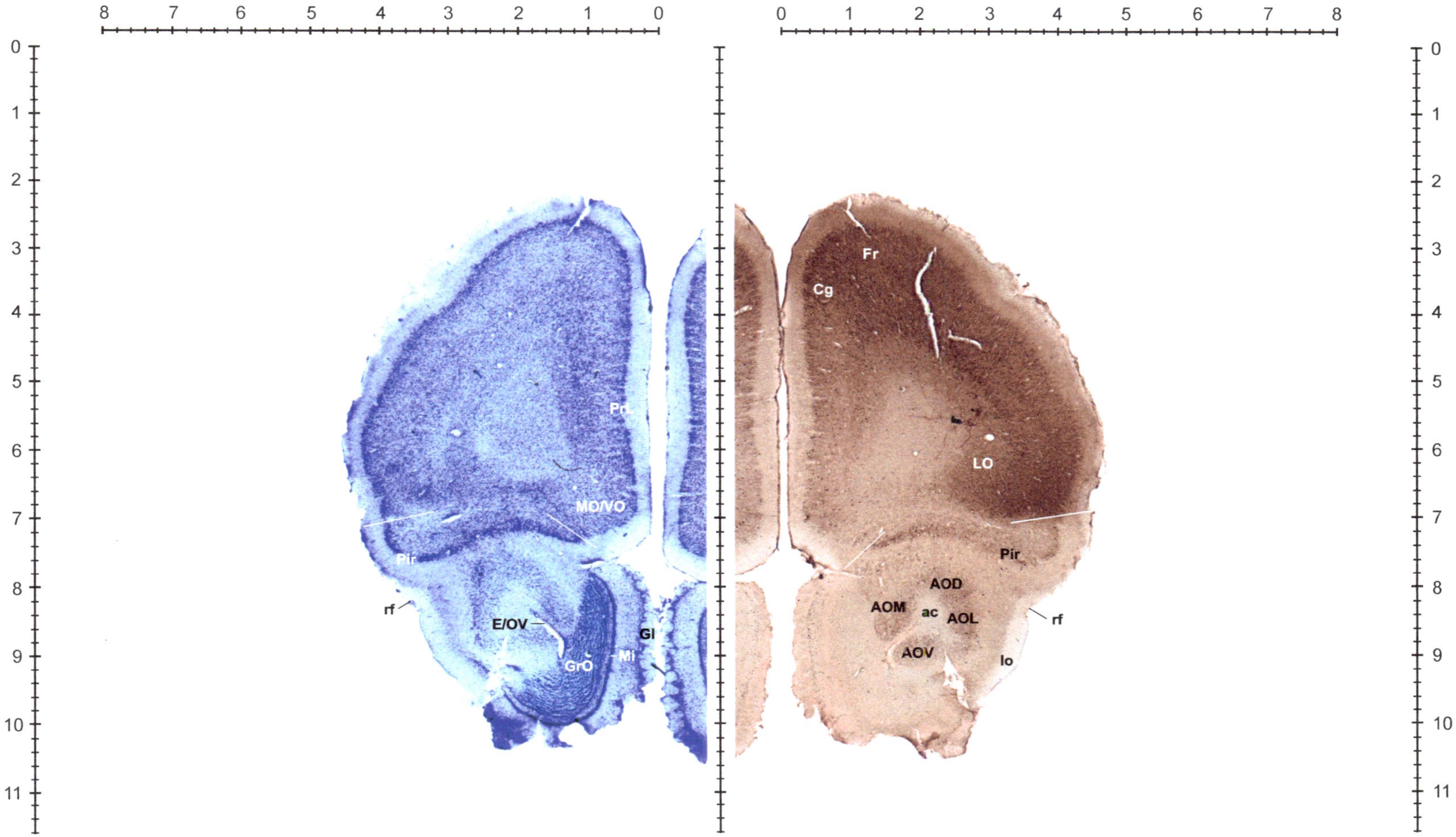


**ac** anterior commissure
**AOD** anterior olfactory nucleus, dorsal part
**AOL** anterior olfactory nucleus, lateral part
**AOM** anterior olfactory nu, medial part
**AOV** anterior olfactory nu, ventral part
**Cg** cingulate cortex
**E** ependymal and subendymal layer
**Fr** frontal cortex
**Gl** glomerular layer of the olfactory bulb
**GrO** granule cell layer of the olfactory bulb
**LO** lateral orbital cortex
**lo** lateral olfactory tract
**Mi** mitral cell layer of the olfactory bulb
**MO** medial orbital cortex
**OV** olfactory ventricle
**Pir** piriform cortex
**PrL** prelimbic cortex
**rf** rhinal fissure
**VO** ventral orbital cortex

**Plate 11 - Sinus: 11.94 mm**

Interaural: 14.64 mm

Nissl section 34

Interaural: 14.60 mm

AChE section 34

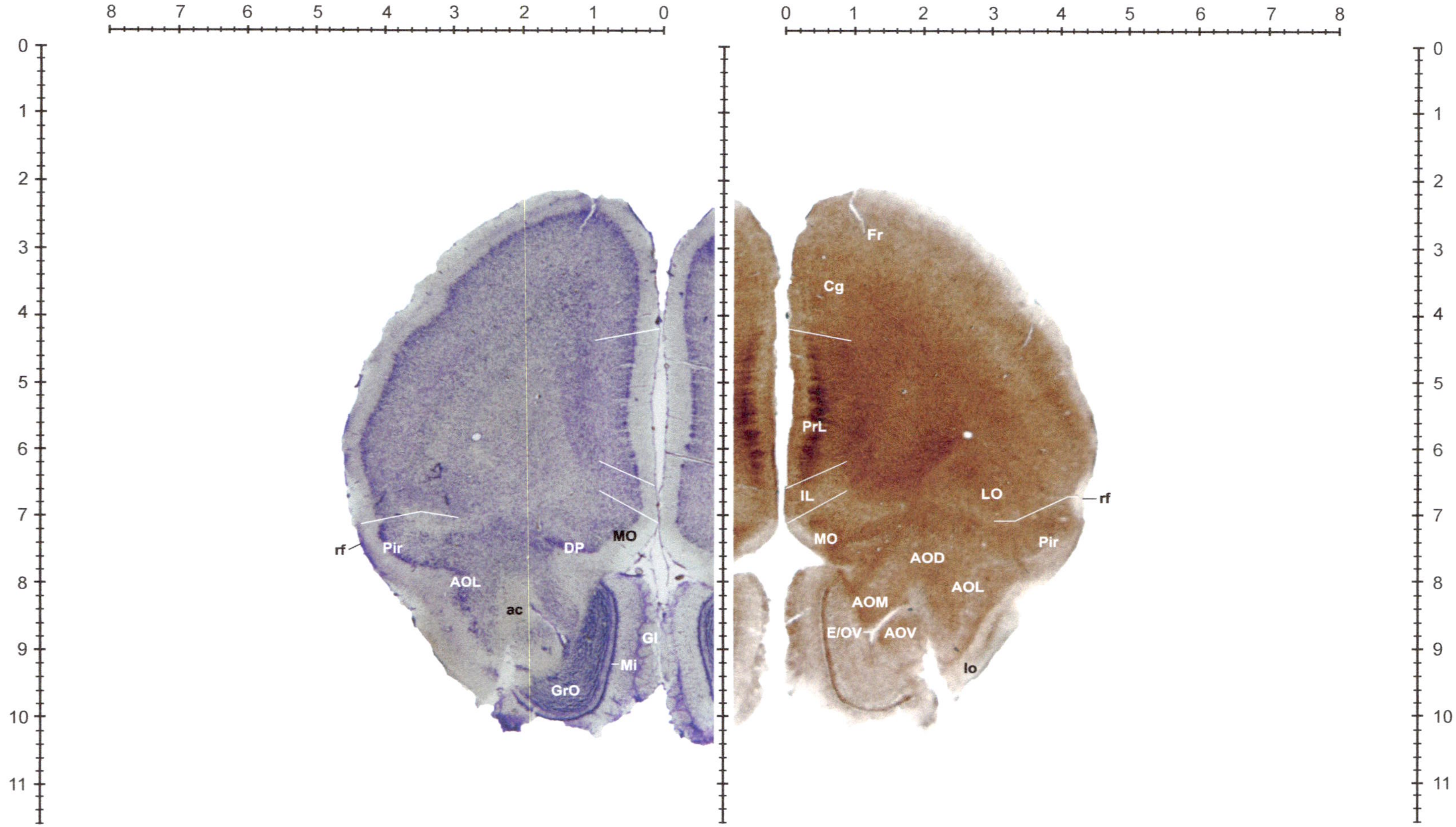


**ac** anterior commissure
**AOD** anterior olfactory nucleus, dorsal part
**AOL** anterior olfactory nucleus, lateral part
**AOM** anterior olfactory nu, medial part
**AOV** anterior olfactory nu, ventral part
**Cg** cingulate cortex
**DP** dorsal peduncular cortex
**E** ependymal and subendymal layer
**Fr** frontal cortex
**Gl** glomerular layer of the olfactory bulb
**GrO** granule cell layer of the olfactory bulb
**IL** infralimbic cortex
**LO** lateral orbital cortex
**lo** lateral olfactory tract
**Mi** mitral cell layer of the olfactory bulb
**MO** medial orbital cortex
**OV** olfactory ventricle
**Pir** piriform cortex
**PrL** prelimbic cortex
**rf** rhinal fissure

## Plate 12 - Sinus: 11.78 mm

Interaural: 14.48 mm
Nissl section 36

Interaural: 14.43 mm
AChE section 36

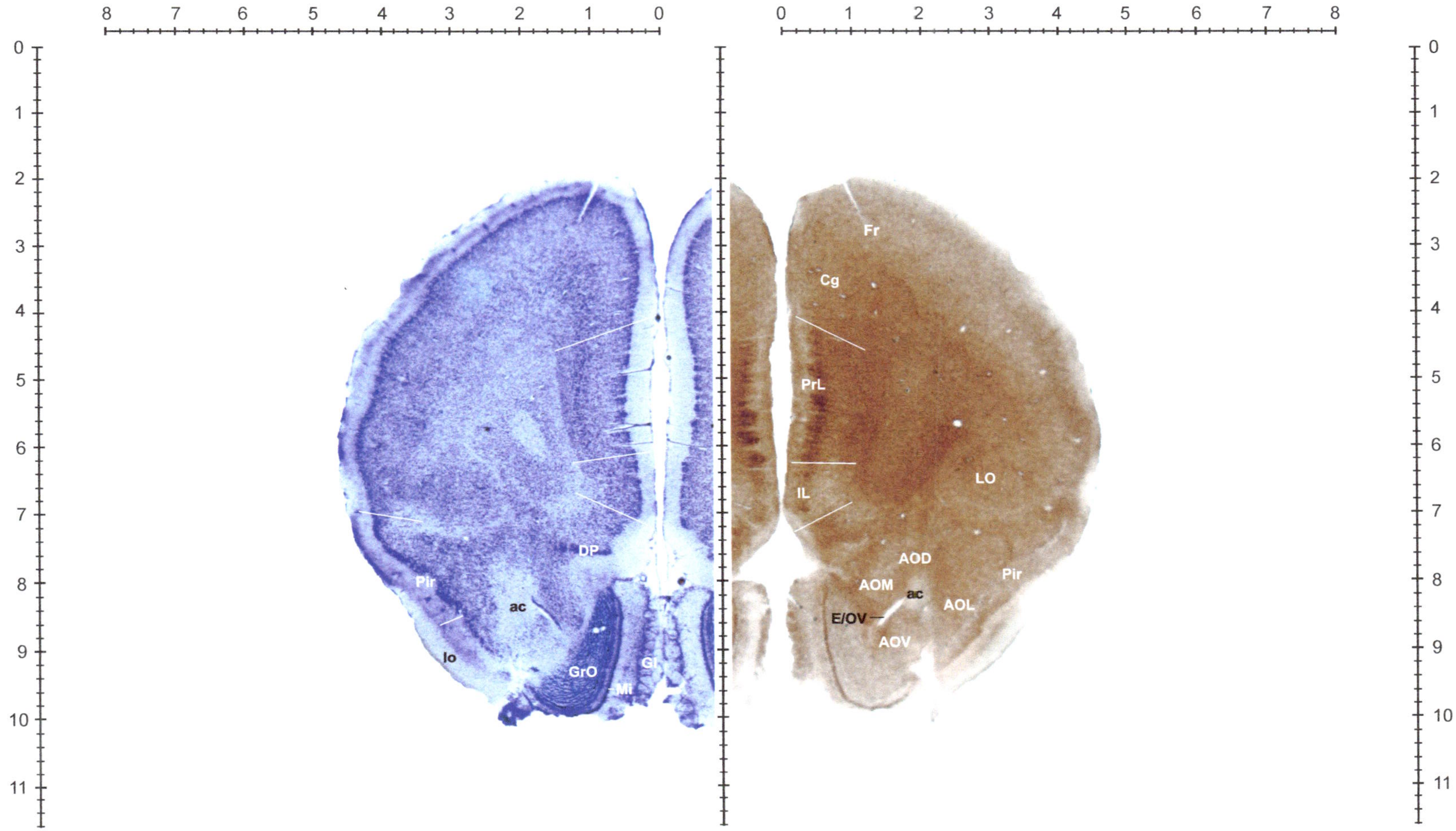


ac anterior commissure
AOD anterior olfactory nucleus, dorsal part
AOL anterior olfactory nucleus, lateral part
AOM anterior olfactory nu, medial part
AOV anterior olfactory nu, ventral part
Cg cingulate cortex
DP dorsal peduncular cortex
E ependymal and subendymal layer
Fr frontal cortex
Gl glomerular layer of the olfactory bulb
GrO granule cell layer of the olfactory bulb
IL infralimbic cortex
LO lateral orbital cortex
lo lateral olfactory tract
Mi mitral cell layer of the olfactory bulb
OV olfactory ventricle
Pir piriform cortex
PrL prelimbic cortex

**Plate 13 - Sinus: 11.61 mm**

Interaural: 14.31 mm
Nissl section 38

Interaural: 14.27 mm
Parv section 38

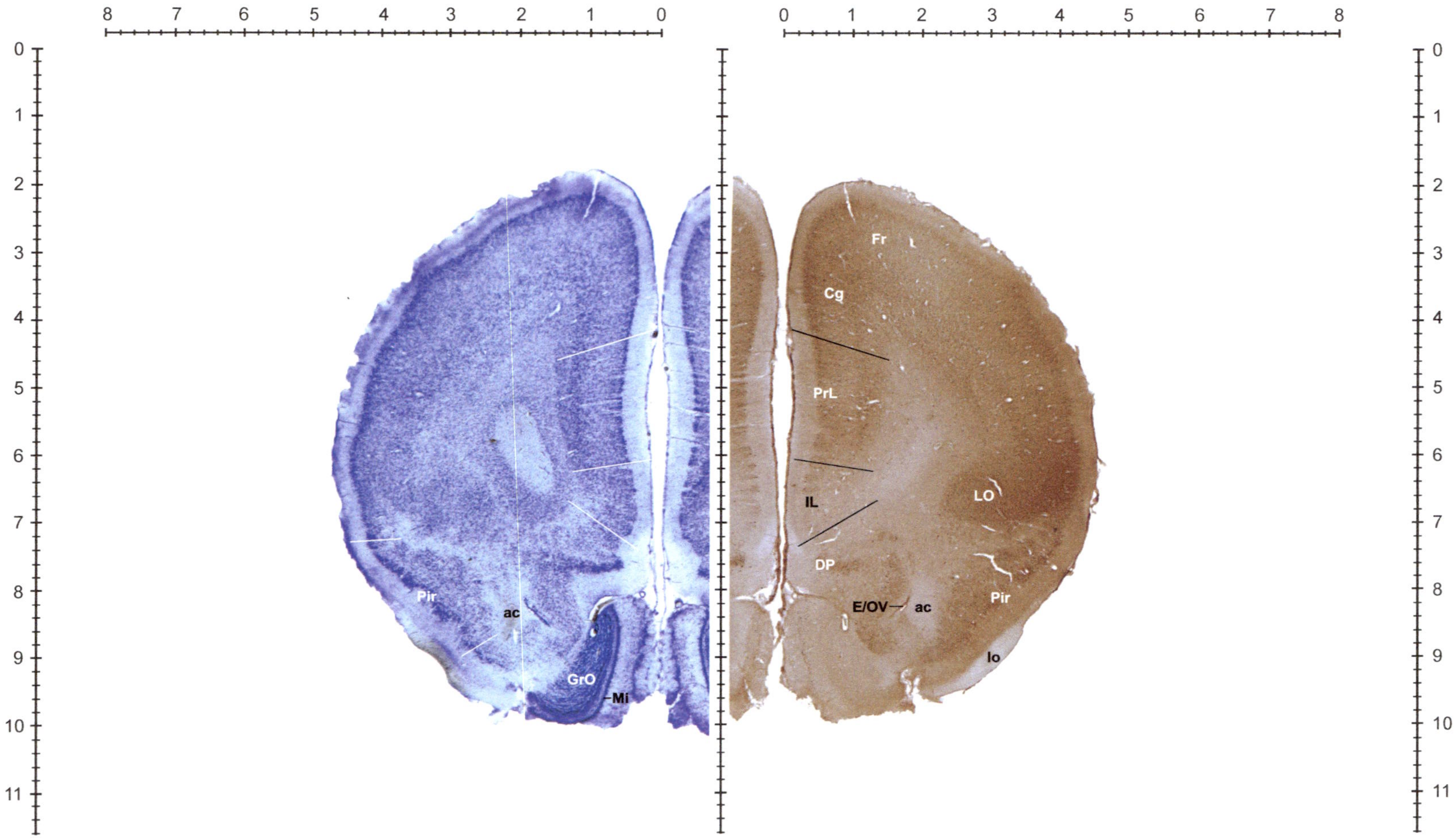


**ac** anterior commissure
**Cg** cingulate cortex
**DP** dorsal peduncular cortex
**E** ependymal and subendymal layer
**Fr** frontal cortex
**GrO** granule cell layer of the olfactory bulb
**IL** infralimbic cortex
**LO** lateral orbital cortex
**lo** lateral olfactory tract
**Mi** mitral cell layer of the olfactory bulb
**OV** olfactory ventricle
**Pir** piriform cortex
**PrL** prelimbic cortex

# Plate 14 - Sinus: 11.53 mm

Interaural: 14.23 mm
Nissl section 39

Interaural: 14.11 mm
AChE section 40

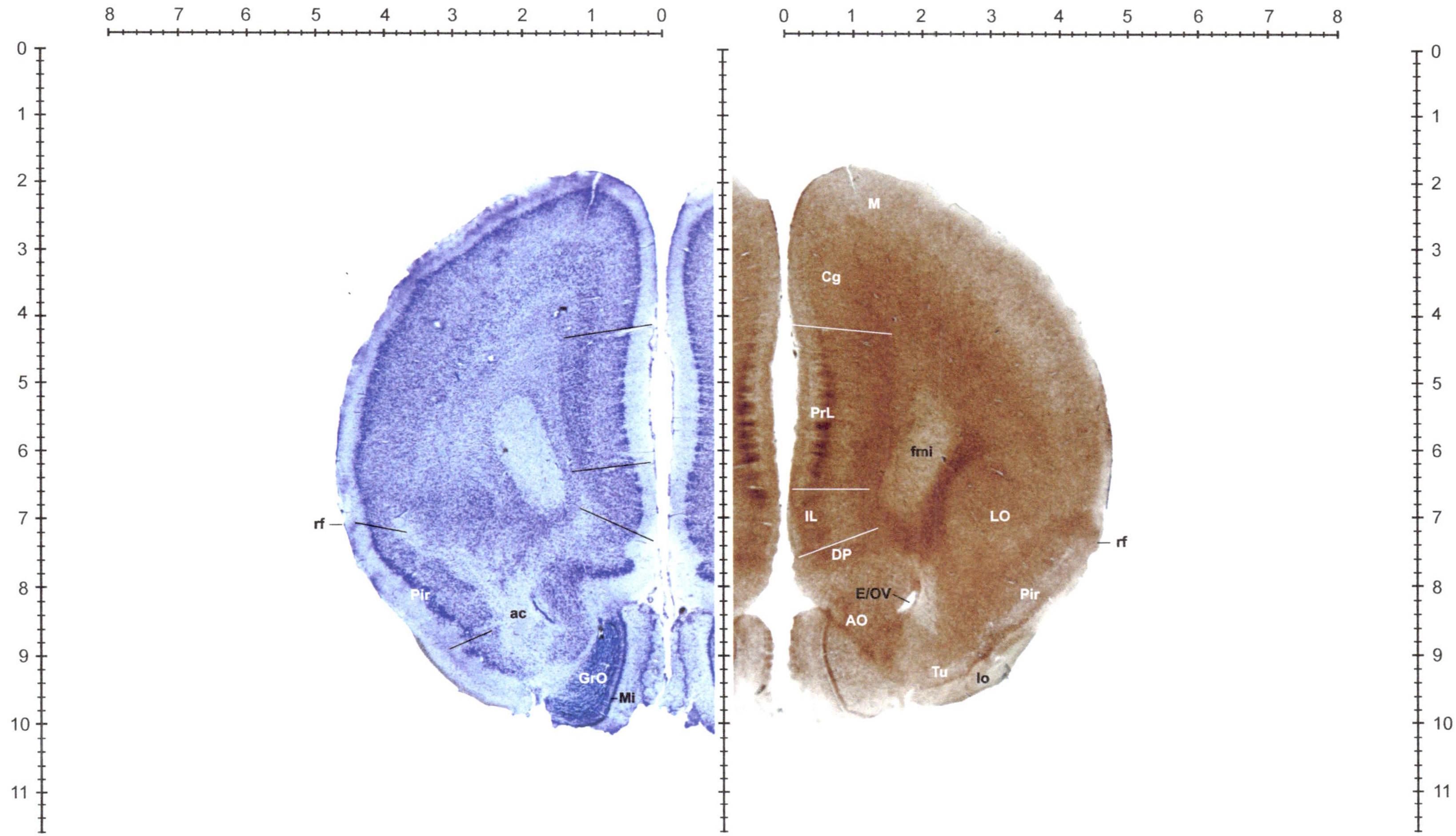


**ac** anterior commissure
**AO** anterior olfactory area
**Cg** cingulate cortex
**DP** dorsal peduncular cortex
**E** ependymal and subendymal layer
**fmi** forceps minor of the corpus callosum
**GrO** granule cell layer of the olfactory bulb
**IL** infralimbic cortex
**LO** lateral orbital cortex
**lo** lateral olfactory tract
**M** motor cortex
**Mi** mitral cell layer of the olfactory bulb
**OV** olfactory ventricle
**Pir** piriform cortex
**PrL** prelimbic cortex
**rf** rhinal fissure
**Tu** olfactory tubercle

**Plate 15 - Sinus: 11.20 mm**

Interaural: 13.90 mm
Nissl section 43

Interaural: 13.86 mm
AChE section 43

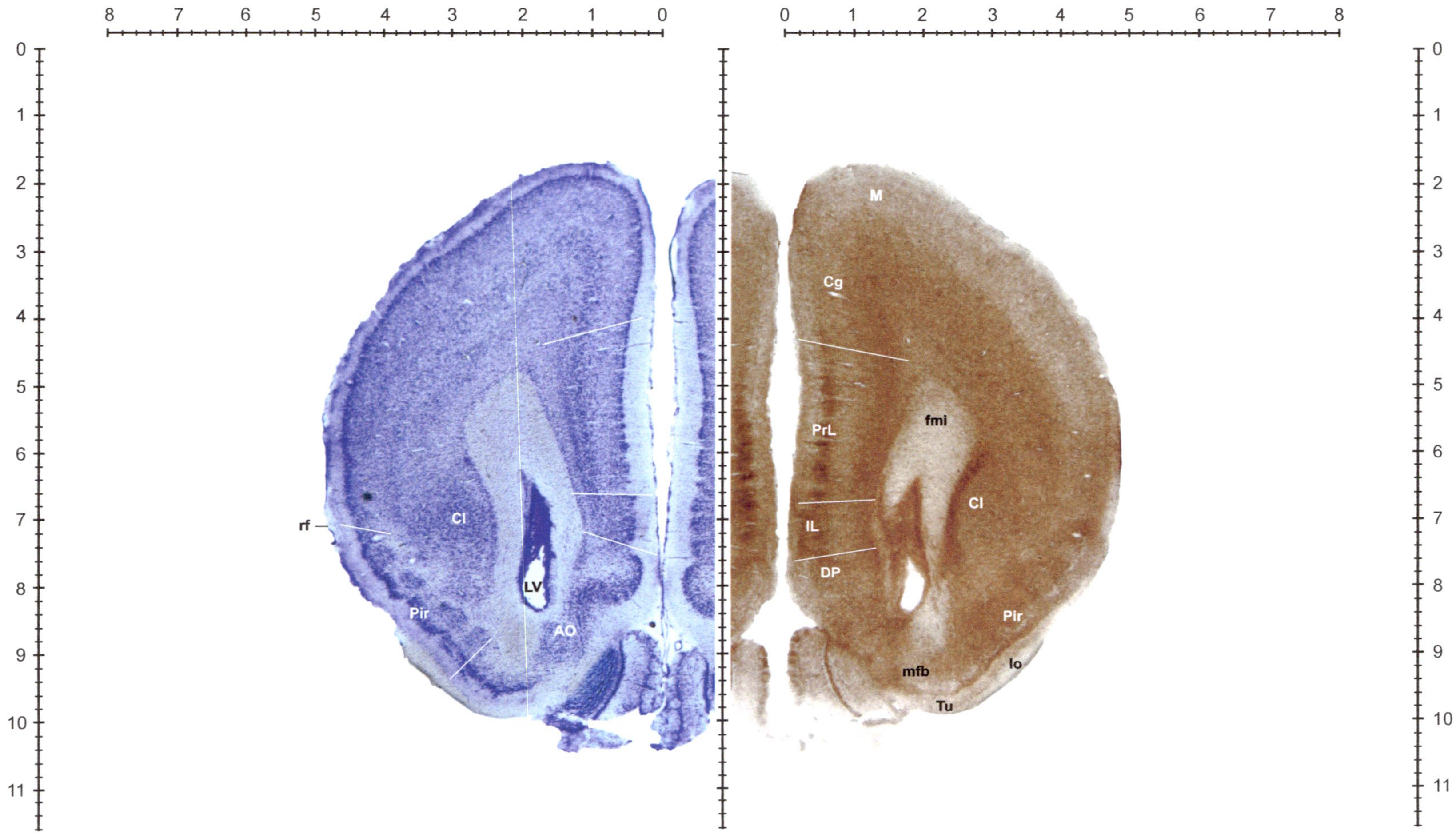


**AO** anterior olfactory area
**Cg** cingulate cortex
**Cl** claustrum
**DP** dorsal peduncular cortex
**fmi** forceps minor of the corpus callosum
**IL** infralimbic cortex
**lo** lateral olfactory tract
**LV** lateral ventricle
**M** motor cortex
**mfb** medial forebrain bundle
**Pir** piriform cortex
**PrL** prelimbic cortex
**rf** rhinal fissure
**Tu** olfactory tubercle

**Plate 16 - Sinus: 11.12 mm**

Interaural: 13.82 mm
Nissl section 44

Interaural: 13.78 mm
Parv section 44

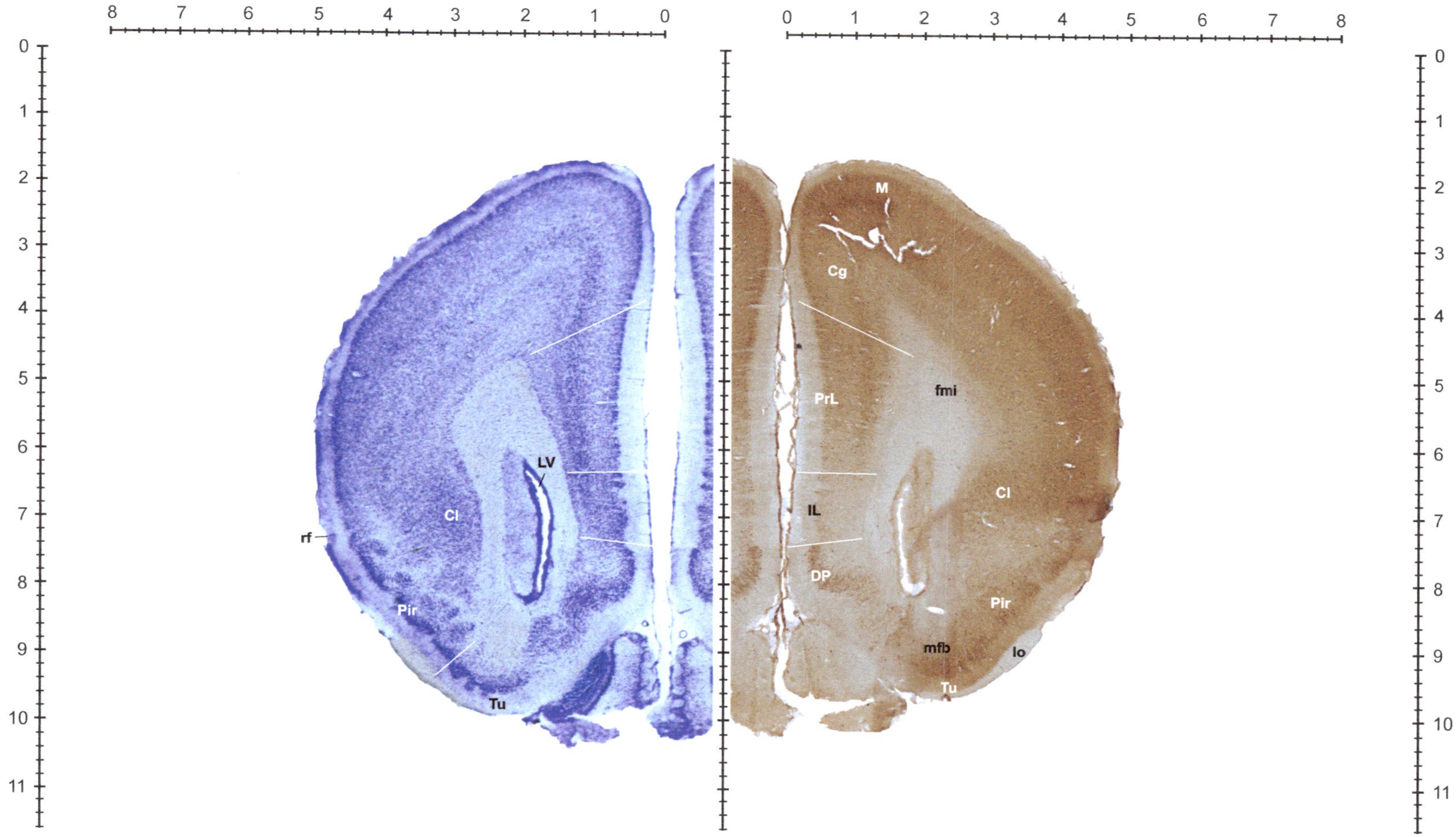


**Cg** cingulate cortex
**Cl** claustrum
**DP** dorsal peduncular cortex
**fmi** forceps minor of the corpus callosum
**IL** infralimbic cortex
**lo** lateral olfactory tract
**LV** lateral ventricle
**M** motor cortex
**mfb** medial forebrain bundle
**Pir** piriform cortex
**PrL** prelimbic cortex
**rf** rhinal fissure
**Tu** olfactory tubercle

## Plate 17 - Sinus: 10.96 mm

Interaural: 13.66 mm
Nissl section 46

Interaural: 13.61 mm
AChE section 46

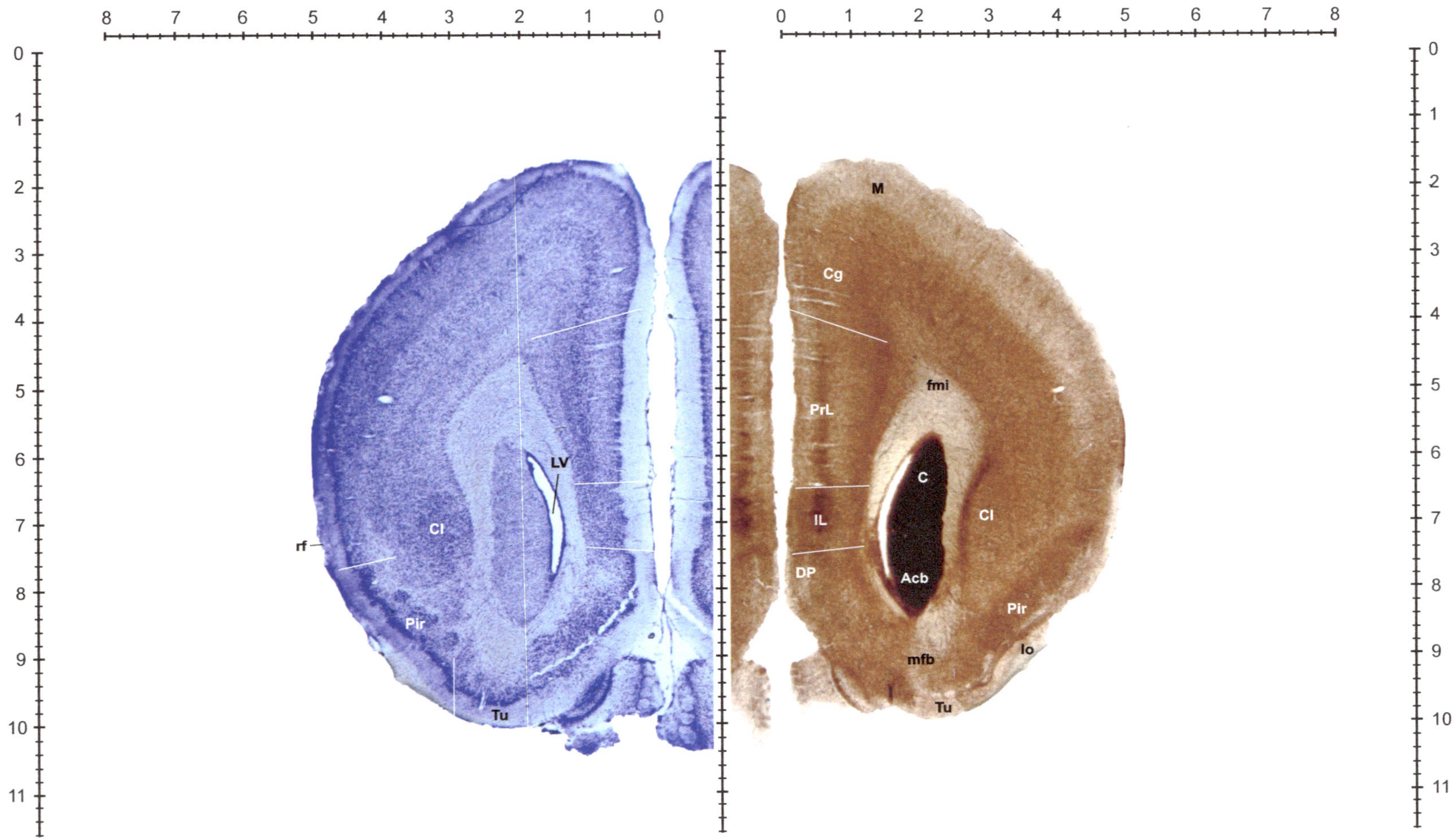


**Acb** accumbens nucleus
**C** caudate
**Cg** cingulate cortex
**Cl** claustrum
**DP** dorsal peduncular cortex
**IL** infralimbic cortex
**fmi** forceps minor of the corpus callosum
**lo** lateral olfactory tract
**LV** lateral ventricle
**M** motor cortex
**mfb** medial forebrain bundle
**Pir** piriform cortex
**PrL** prelimbic cortex
**rf** rhinal fissure
**Tu** olfactory tubercle

## Plate 18 - Sinus: 10.79 mm

Interaural: 13.49 mm
Nissl section 48

Interaural: 13.45 mm
AChE section 48

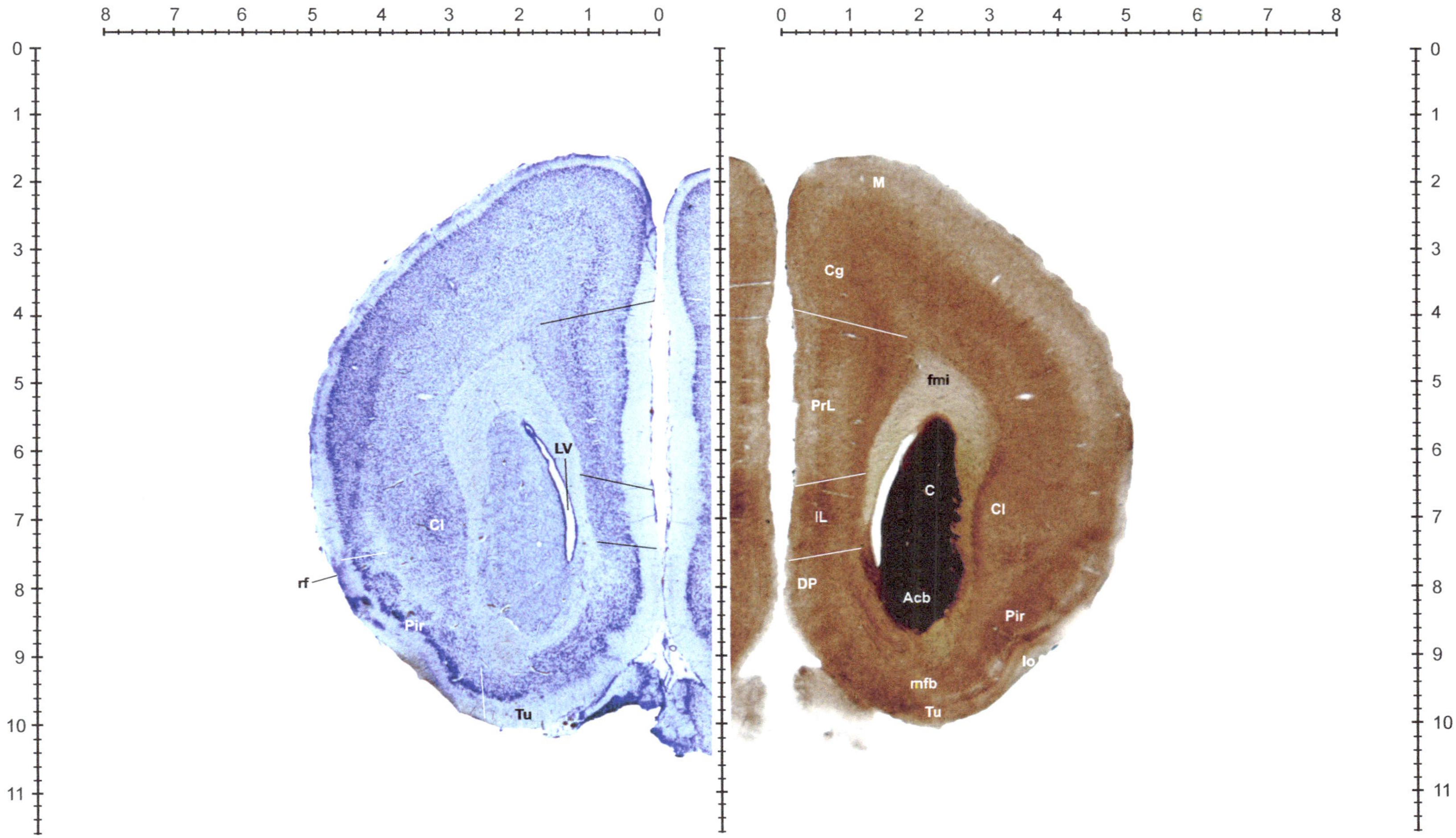


**Acb** accumbens nucleus
**C** caudate
**Cg** cingulate cortex
**Cl** claustrum
**DP** dorsal peduncular cortex
**fmi** forceps minor of the corpus callosum
**IL** infralimbic cortex
**lo** lateral olfactory tract
**LV** lateral ventricle
**M** motor cortex
**mfb** medial forebrain bundle
**Pir** piriform cortex
**PrL** prelimbic cortex
**rf** rhinal fissure
**Tu** olfactory tubercle

## Plate 19 - Sinus: 10.63 mm

Interaural: 13.33 mm
Nissl section 50

Interaural: 13.29 mm
Parv section 50

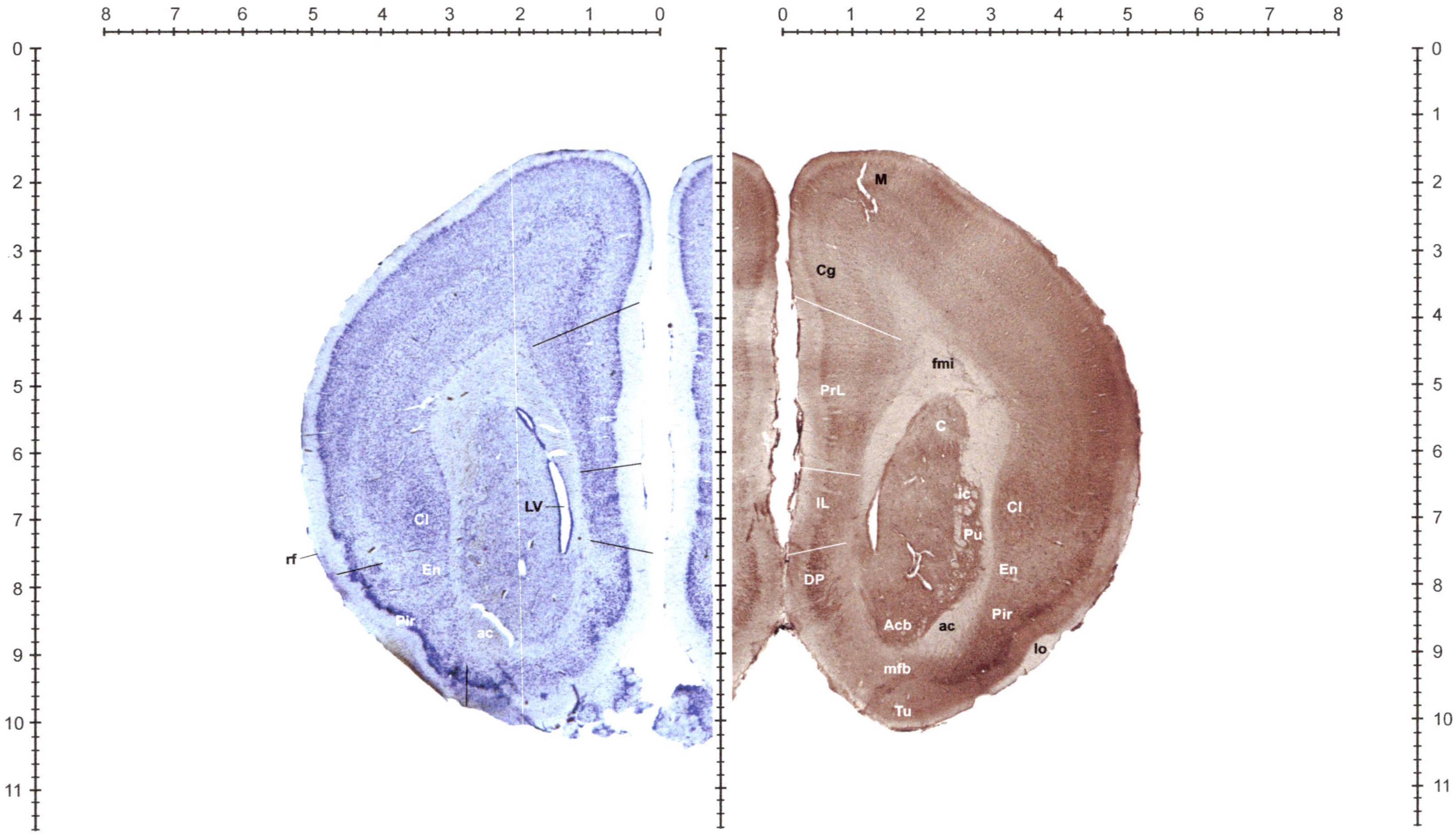


**ac** anterior commissure
**Acb** accumbens nucleus
**C** caudate
**Cg** cingulate cortex
**Cl** claustrum
**DP** dorsal peduncular cortex
**En** endopiriform nucleus
**fmi** forceps minor of the corpus callosum
**ic** internal capsule
**IL** infralimbic cortex
**lo** lateral olfactory tract
**LV** lateral ventricle
**M** motor cortex
**mfb** medial forebrain bundle
**Pir** piriform cortex
**PrL** prelimbic cortex
**Pu** putamen
**rf** rhinal fissure
**Tu** olfactory tubercle

## Plate 20 - Sinus: 10.46 mm

Interaural: 13.16 mm
Nissl section 52

Interaural: 13.20 mm
AChE section 51

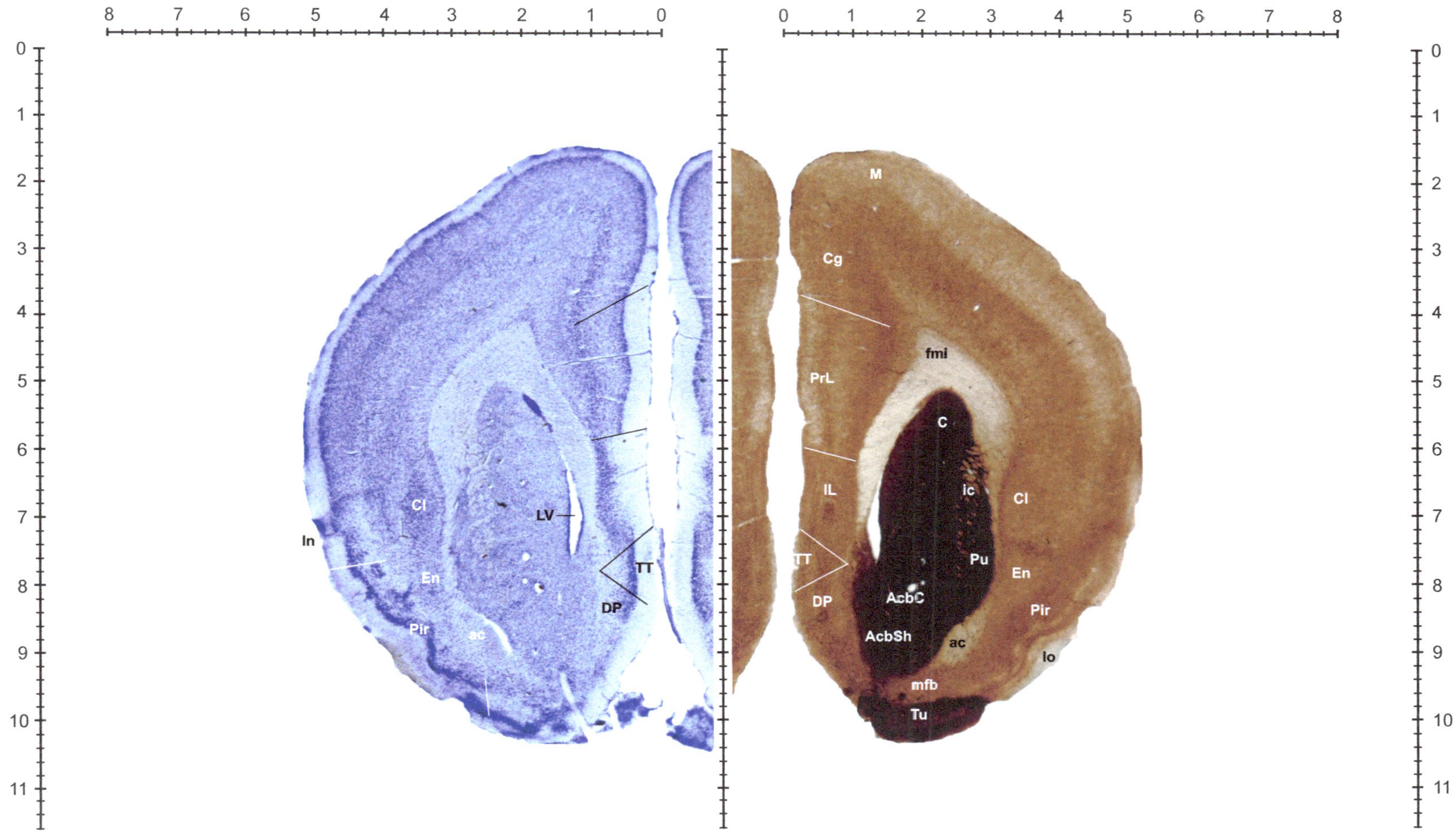


**ac** anterior commissure
**AcbC** accumbens nucleus, core
**AcbSh** accumbens nucleus, shell
**C** caudate
**Cg** cingulate cortex
**Cl** claustrum
**DP** dorsal peduncular cortex
**En** endopiriform nucleus
**fmi** forceps minor of the corpus callosum
**ic** internal capsule
**IL** infralimbic cortex
**In** insular cortex
**lo** lateral olfactory tract
**LV** lateral ventricle
**M** motor cortex
**mfb** medial forebrain bundle
**Pir** piriform cortex
**PrL** prelimbic cortex
**Pu** putamen
**TT** tenia tecta
**Tu** olfactory tubercle

## Plate 21 - Sinus: 10.30 mm

Interaural: 13.00 mm
Nissl section 54

Interaural: 12.96 mm
AChE section 54

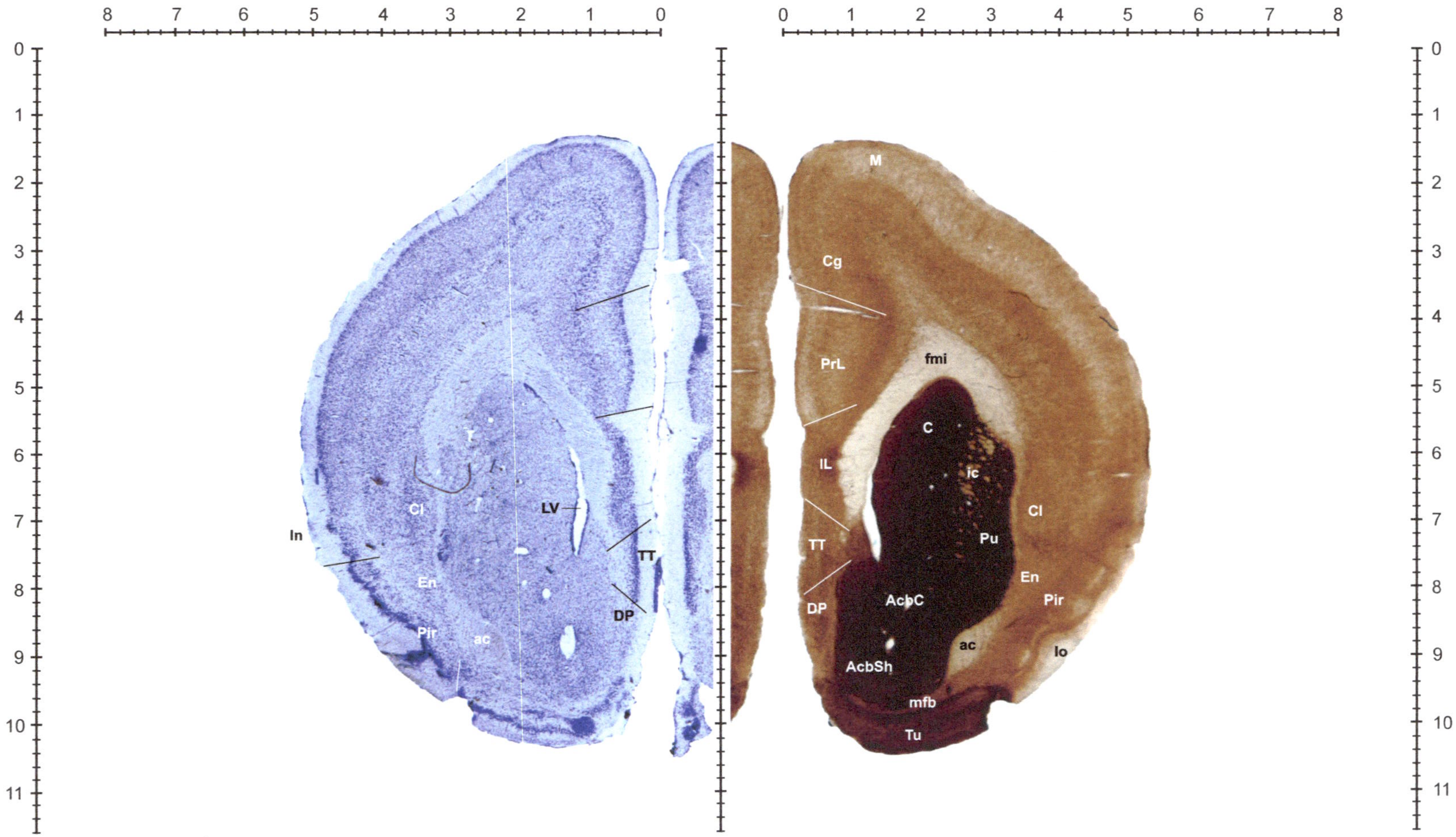


**ac** anterior commissure
**AcbC** accumbens nucleus, core
**AcbSh** accumbens nucleus, shell
**C** caudate
**Cg** cingulate cortex
**Cl** claustrum
**DP** dorsal peduncular cortex
**En** endopiriform nucleus
**fmi** forceps minor of the corpus callosum
**ic** internal capsule
**IL** infralimbic cortex
**In** insular cortex
**lo** lateral olfactory tract
**LV** lateral ventricle
**M** motor cortex
**mfb** medial forebrain bundle
**Pir** piriform cortex
**PrL** prelimbic cortex
**Pu** putamen
**Tu** olfactory tubercle
**TT** tenia tecta

## Plate 22 - Sinus: 10.22 mm

Interaural: 12.92 mm
Nissl section 55

Interaural: 12.79 mm
Parv section 56

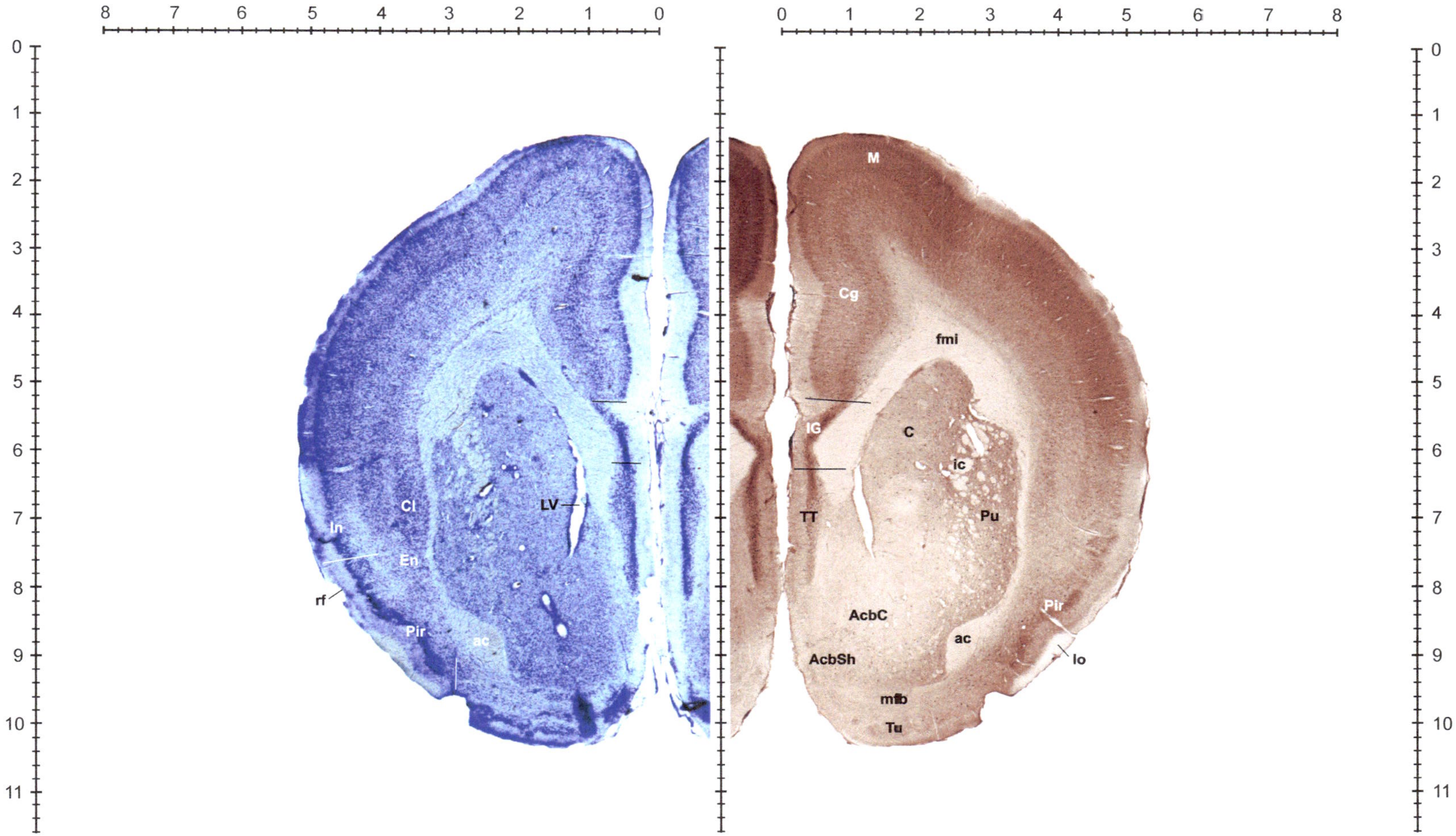


**ac** anterior commissure
**AcbC** accumbens nucleus, core
**AcbSh** accumbens nucleus, shell
**C** caudate
**Cg** cingulate cortex
**Cl** claustrum
**En** endopiriform nucleus
**fmi** forceps minor of the corpus callosum
**IG** indusium griseum
**ic** internal capsule
**In** insular cortex
**lo** lateral olfactory tract
**LV** lateral ventricle
**M** motor cortex
**mfb** medial forebrain bundle
**Pir** piriform cortex
**Pu** putamen
**rf** rhinal fissure
**Tu** olfactory tubercle
**TT** tenia tecta

**Plate 23 - Sinus: 10.05 mm**

Interaural: 12.75 mm
Nissl section 57

Interaural: 12.71 mm
AChE section 57

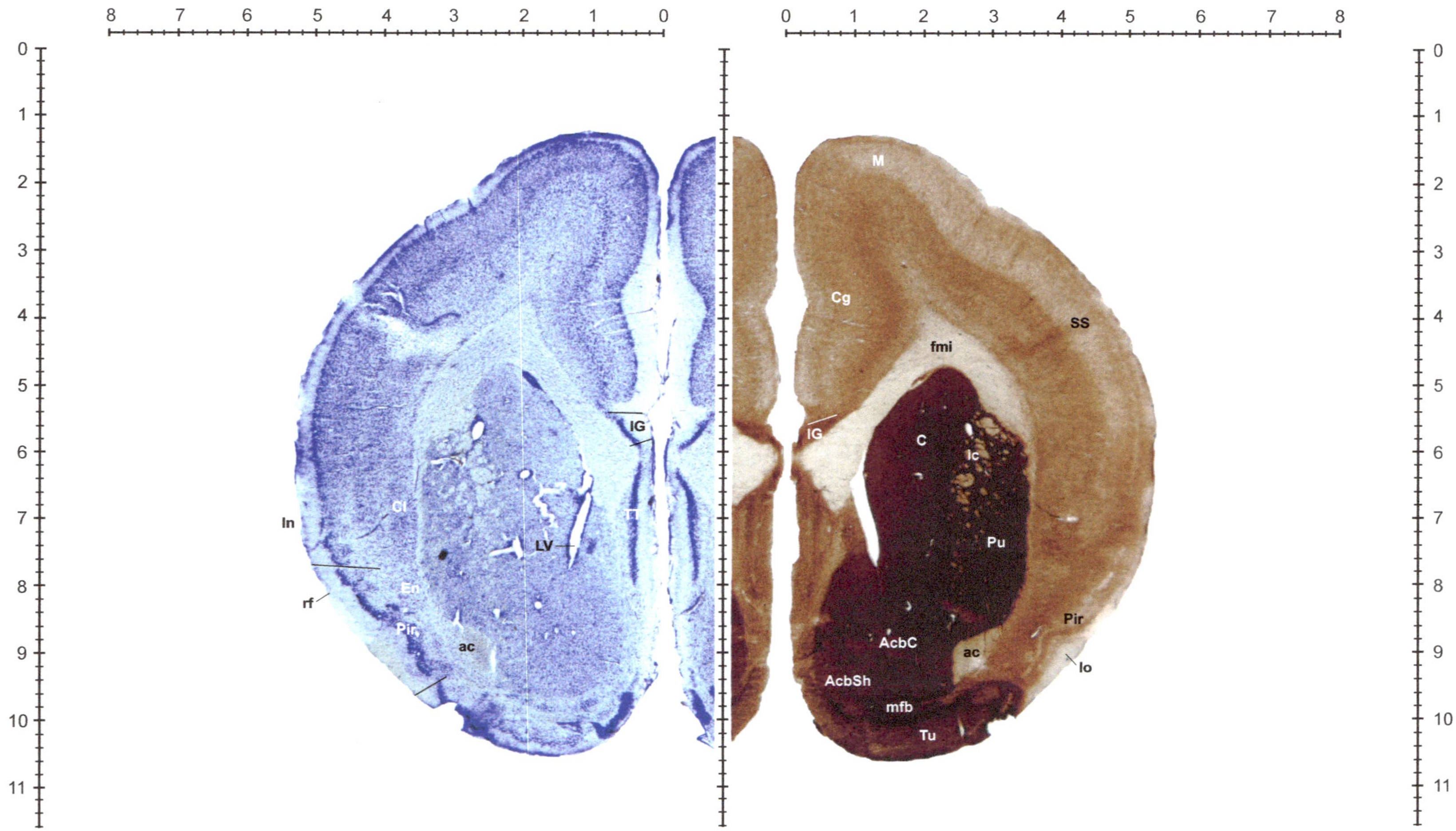


**ac** anterior commissure
**AcbC** accumbens nucleus, core
**AcbSh** accumbens nucleus, shell
**C** caudate
**Cg** cingulate cortex
**Cl** claustrum
**En** endopiriform nucleus
**fmi** forceps minor of the corpus callosum
**ic** internal capsule
**IG** indusium griseum
**In** insular cortex
**lo** lateral olfactory tract
**LV** lateral ventricle
**M** motor cortex
**mfb** medial forebrain bundle
**Pir** piriform cortex
**Pu** putamen
**rf** rhinal fissure
**SS** somatosensory cortex
**Tu** olfactory tubercle
**TT** tenia tecta

## Plate 24 - Sinus: 9.81 mm

Interaural: 12.51 mm
Nissl section 60

Interaural: 12.47 mm
AChE section 60

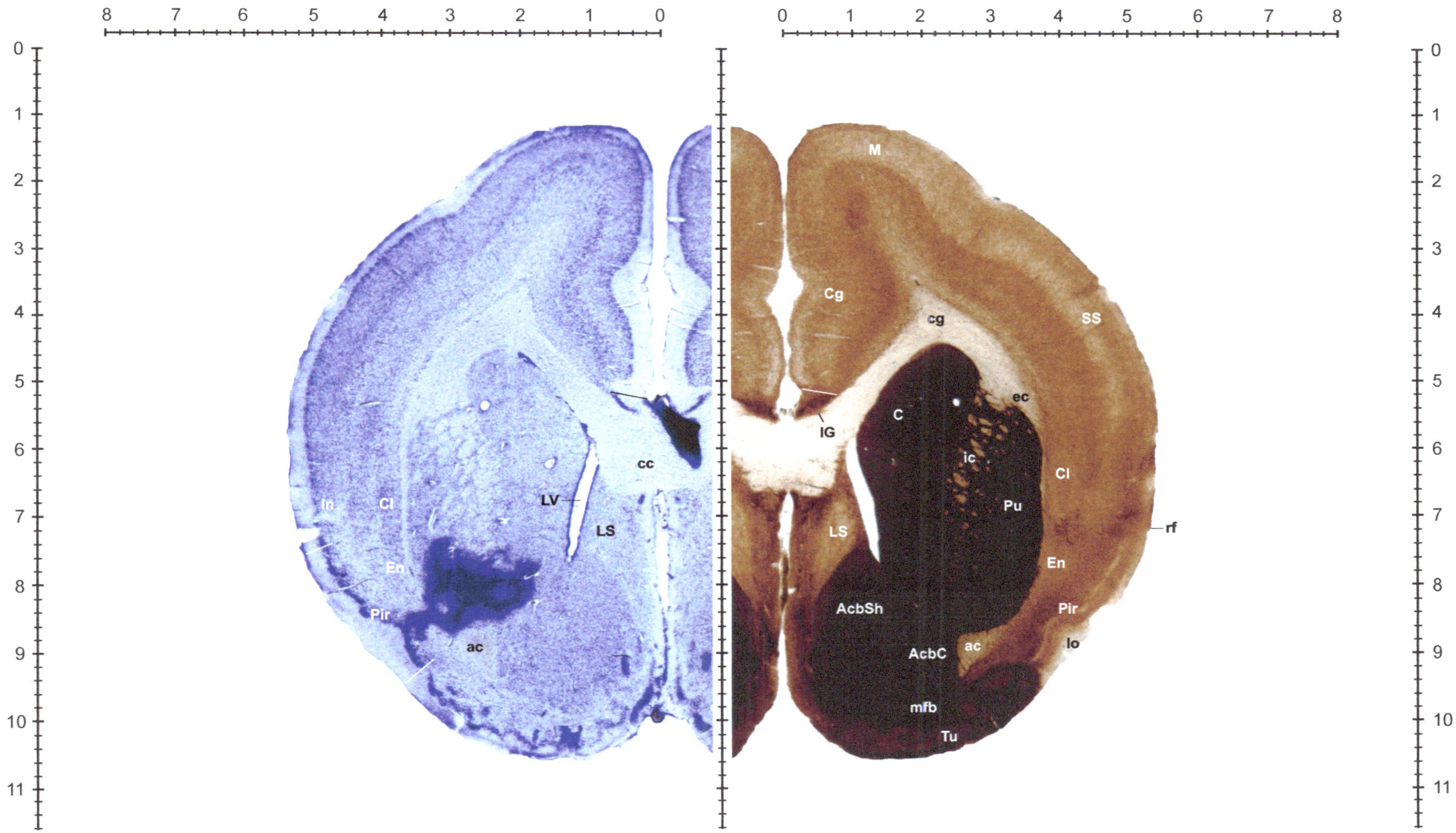


**ac** anterior commissure
**AcbC** accumbens nucleus, core
**AcbSh** accumbens nucleus, shell
**C** caudate
**cc** corpus callosum
**Cg** cingulate cortex
**cg** cingulum
**Cl** claustrum
**ec** external capsule
**En** endopiriform nucleus
**ic** internal capsule
**IG** indusium griseum
**In** insular cortex
**lo** lateral olfactory tract
**LS** lateral septal nucleus
**LV** lateral ventricle
**M** motor cortex
**mfb** medial forebrain bundle
**Pir** piriform cortex
**Pu** putamen
**rf** rhinal fissure
**SS** somatosensory cortex
**Tu** olfactory tubercle

## Plate 25 - Sinus: 9.64 mm

Interaural: 12.34 mm
Nissl section 62

Interaural: 12.30 mm
Parv section 62

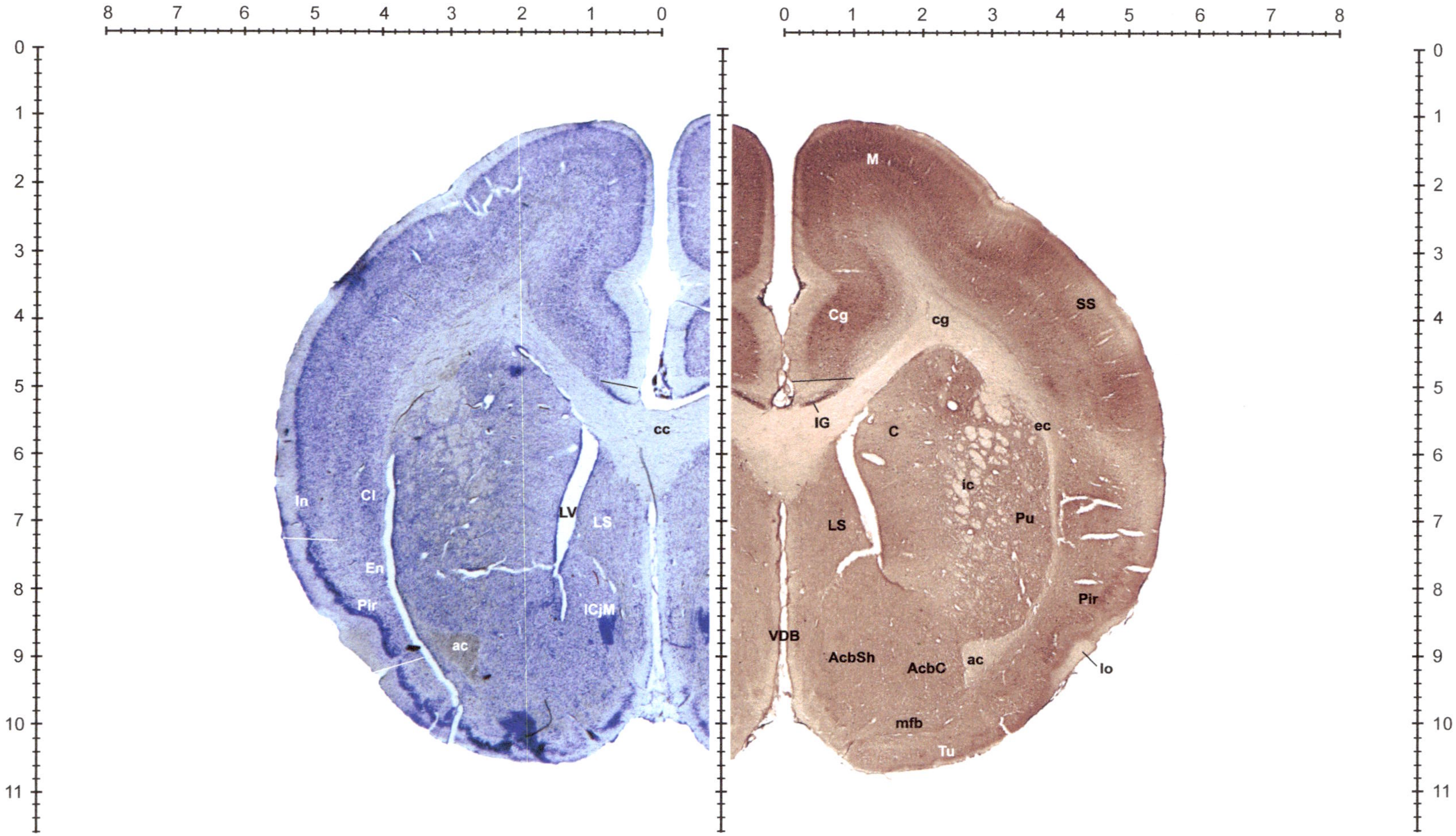


**ac** anterior commissure
**AcbC** accumbens nucleus, core
**AcbSh** accumbens nucleus, shell
**C** caudate
**cc** corpus callosum
**Cg** cingulate cortex
**cg** cingulum
**Cl** claustrum
**ec** external capsule
**En** endopiriform nucleus
**ic** internal capsule
**ICjM** island of Calleja, major island
**IG** indusium griseum
**In** insular cortex
**lo** lateral olfactory tract
**LS** lateral septal nucleus
**LV** lateral ventricle
**M** motor cortex
**mfb** medial forebrain bundle
**Pir** piriform cortex
**Pu** putamen
**SS** somatosensory cortex
**Tu** olfactory tubercle
**VDB** vertical limb nu diagonal band

## Plate 26 - Sinus: 9.40 mm

Interaural: 12.10 mm
Nissl section 65

Interaural: 12.14 mm
AChE section 64

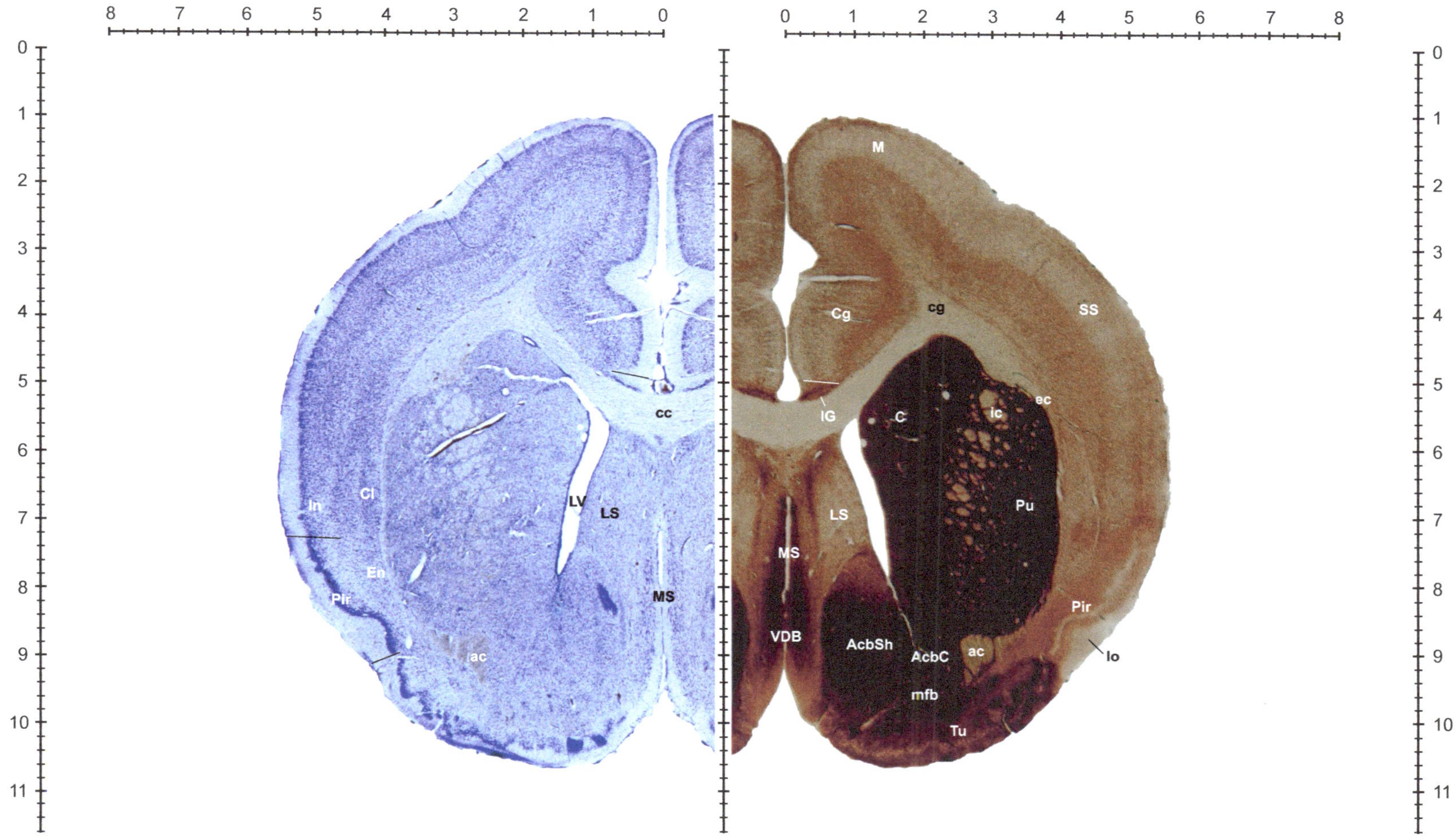


**ac** anterior commissure
**AcbC** accumbens nucleus, core
**AcbSh** accumbens nucleus, shell
**C** caudate
**cc** corpus callosum
**Cg** cingulate cortex
**cg** cingulum
**Cl** claustrum
**ec** external capsule
**En** endopiriform nucleus
**ic** internal capsule
**IG** indusium griseum
**In** insular cortex
**lo** lateral olfactory tract
**LS** lateral septal nucleus
**LV** lateral ventricle
**M** motor cortex
**mfb** medial forebrain bundle
**MS** medial septal nucleus
**Pir** piriform cortex
**Pu** putamen
**SS** somatosensory cortex
**Tu** olfactory tubercle
**VDB** vertical limb nu diagonal band

**Plate 27 - Sinus: 9.32 mm**

Interaural: 12.02 mm
Nissl section 66

Interaural: 11.89 mm
AChE section 67

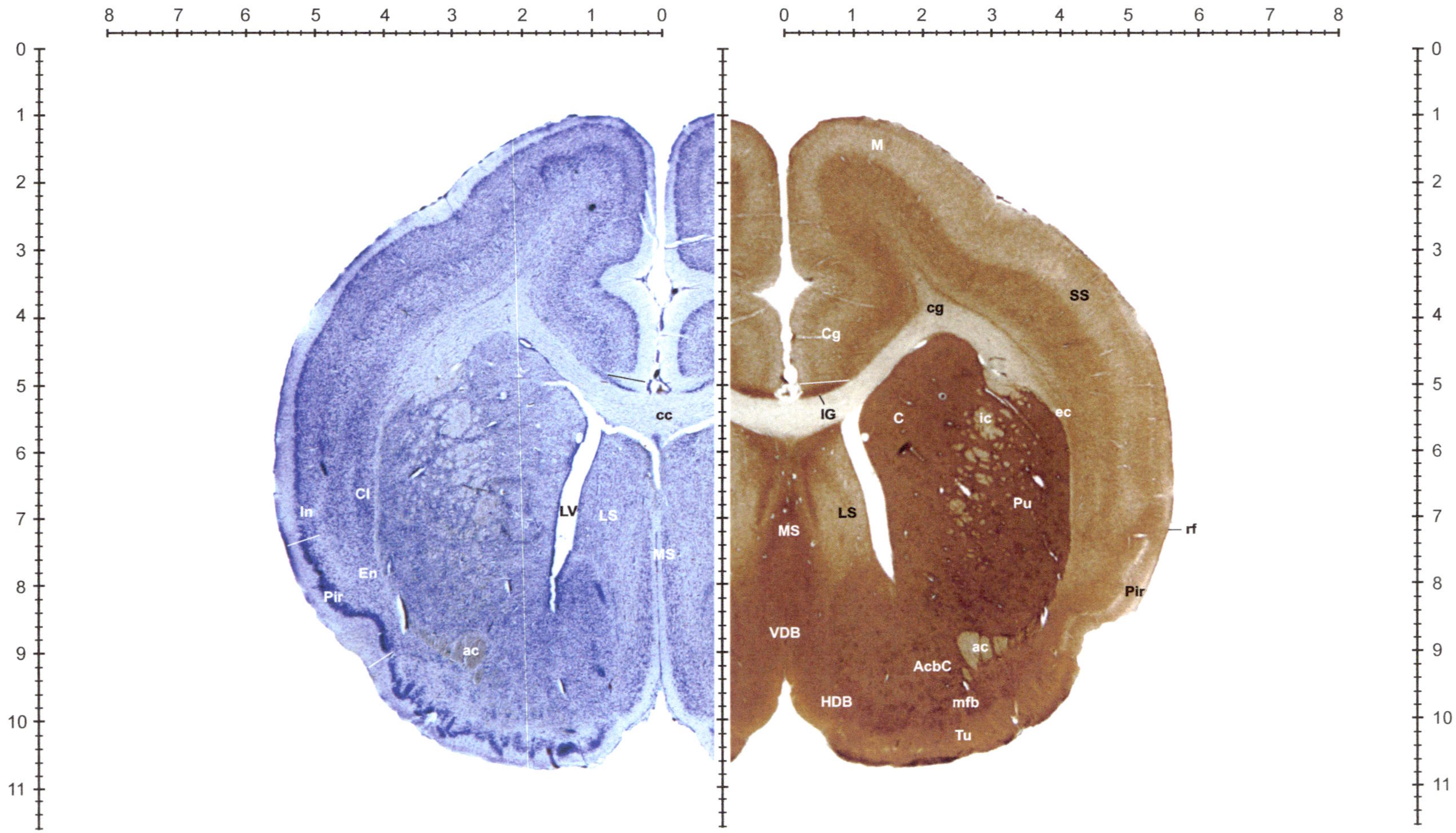


**ac** anterior commissure
**AcbC** accumbens nucleus, core
**C** caudate
**cc** corpus callosum
**Cg** cingulate cortex
**cg** cingulum
**Cl** claustrum
**ec** external capsule
**En** endopiriform nucleus
**HDB** horizontal limb nu diagonal band
**ic** internal capsule
**IG** indusium griseum
**In** insular cortex
**LS** lateral septal nucleus
**LV** lateral ventricle
**M** motor cortex
**mfb** medial forebrain bundle
**MS** medial septal nucleus
**Pir** piriform cortex
**Pu** putamen
**rf** rhinal fissure
**SS** somatosensory cortex
**Tu** olfactory tubercle
**VDB** vertical limb nu diagonal band

## Plate 28 - Sinus: 9.15 mm

Interaural: 11.85 mm
Nissl section 68

Interaural: 11.81 mm
Parv section 68

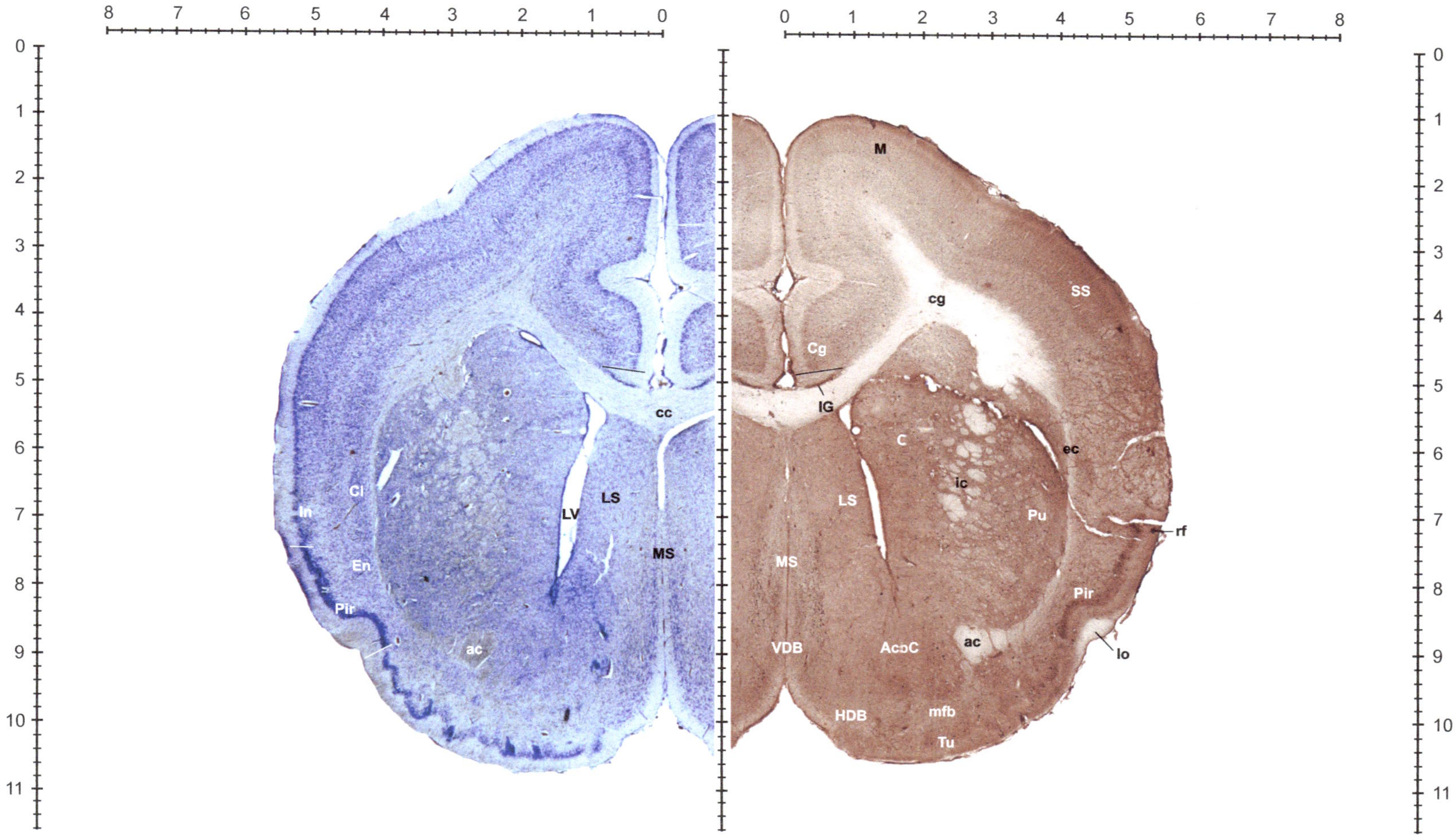


**ac** anterior commissure
**AcbC** accumbens nucleus, core
**C** caudate
**cc** corpus callosum
**Cg** cingulate cortex
**cg** cingulum
**Cl** claustrum
**ec** external capsule
**En** endopiriform nucleus
**HDB** horizontal limb nu diagonal band
**ic** internal capsule
**IG** indusium griseum
**In** insular cortex
**lo** lateral olfactory tract
**LS** lateral septal nucleus
**LV** lateral ventricle
**M** motor cortex
**mfb** medial forebrain bundle
**MS** medial septal nucleus
**Pir** piriform cortex
**Pu** putamen
**rf** rhinal fissure
**SS** somatosensory cortex
**Tu** olfactory tubercle
**VDB** vertical limb nu diagonal band

**Plate 29 - Sinus: 8.99 mm**

Interaural: 11.69 mm
Nissl section 70

Interaural: 11.73 mm
AChE section 69

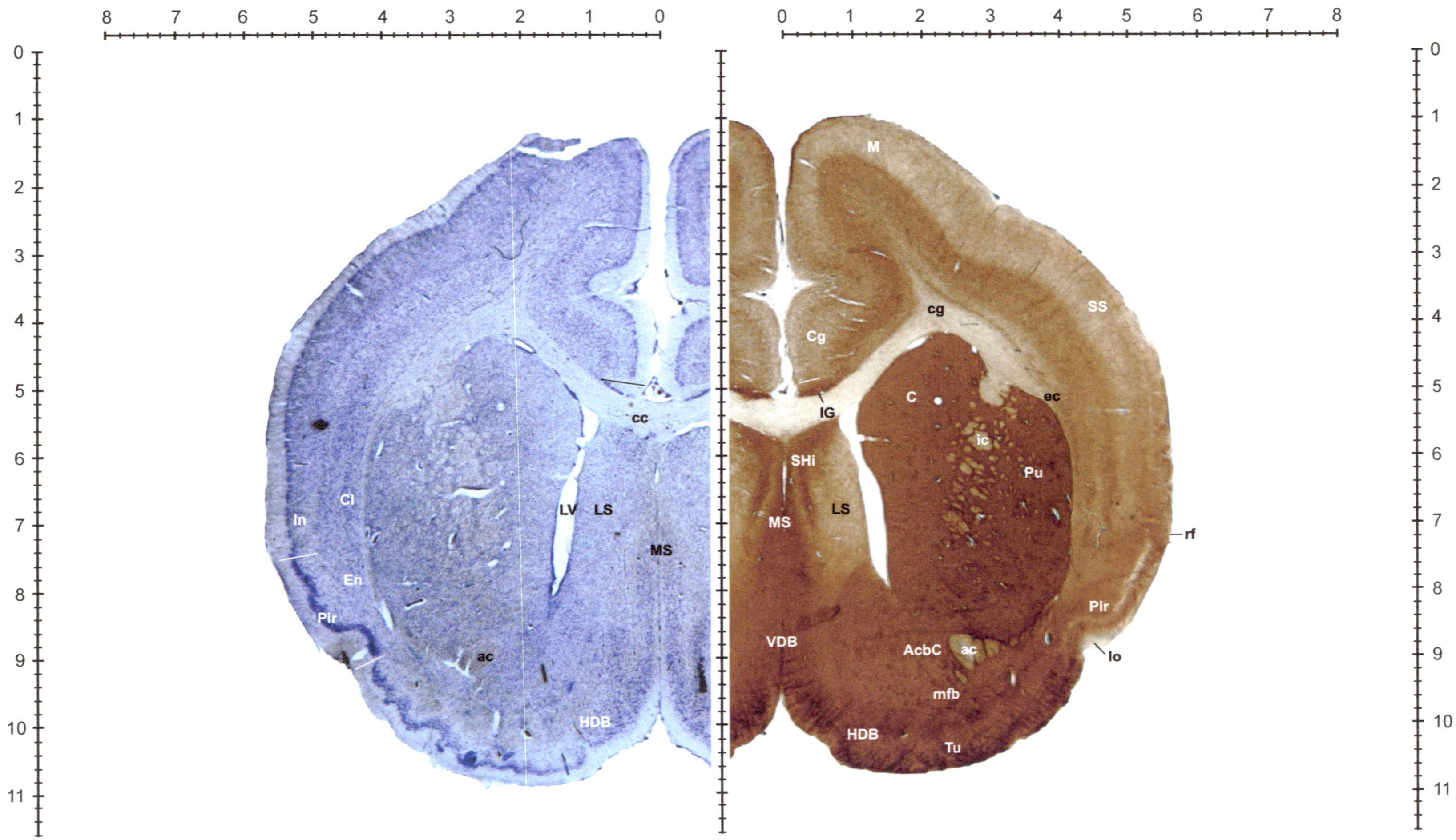


**ac** anterior commissure
**AcbC** accumbens nucleus, core
**C** caudate
**cc** corpus callosum
**Cg** cingulate cortex
**cg** cingulum
**Cl** claustrum
**ec** external capsule
**En** endopiriform nucleus
**HDB** horizontal limb nu diagonal band
**ic** internal capsule
**IG** indusium griseum
**In** insular cortex
**lo** lateral olfactory tract
**LS** lateral septal nucleus
**LV** lateral ventricle
**M** motor cortex
**mfb** medial forebrain bundle
**MS** medial septal nucleus
**Pir** piriform cortex
**Pu** putamen
**rf** rhinal fissure
**SHi** septohippocampal nucleus
**SS** somatosensory cortex
**Tu** olfactory tubercle
**VDB** vertical limb nu diagonal band

## Plate 30 - Sinus: 8.82 mm

Interaural: 11.52 mm
Nissl section 72

Interaural: 11.48 mm
AChE section 72

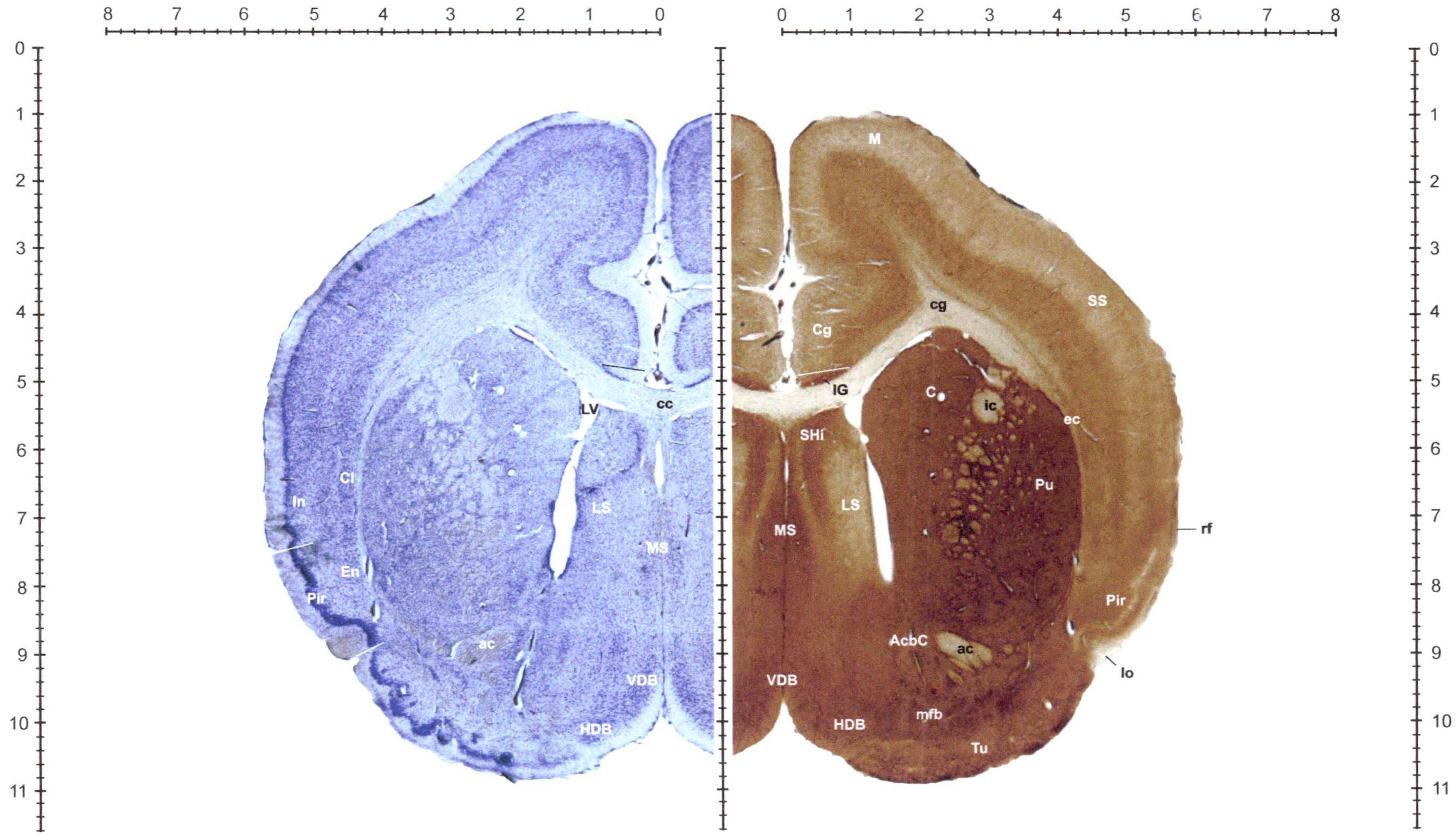


**ac** anterior commissure
**AcbC** accumbens nucleus, core
**C** caudate
**cc** corpus callosum
**Cg** cingulate cortex
**cg** cingulum
**Cl** claustrum
**ec** external capsule
**En** endopiriform nucleus
**HDB** horizontal limb nu diagonal band
**ic** internal capsule
**IG** indusium griseum
**In** insular cortex
**lo** lateral olfactory tract
**LS** lateral septal nucleus
**LV** lateral ventricle
**M** motor cortex
**mfb** medial forebrain bundle
**MS** medial septal nucleus
**Pir** piriform cortex
**Pu** putamen
**rf** rhinal fissure
**SHi** septohippocampal nucleus
**SS** somatosensory cortex
**Tu** olfactory tubercle
**VDB** vertical limb nu diagonal band

## Plate 31 - Sinus: 8.66 mm

Interaural: 11.36 mm
Nissl section 74

Interaural: 11.32 mm
Parv section 74

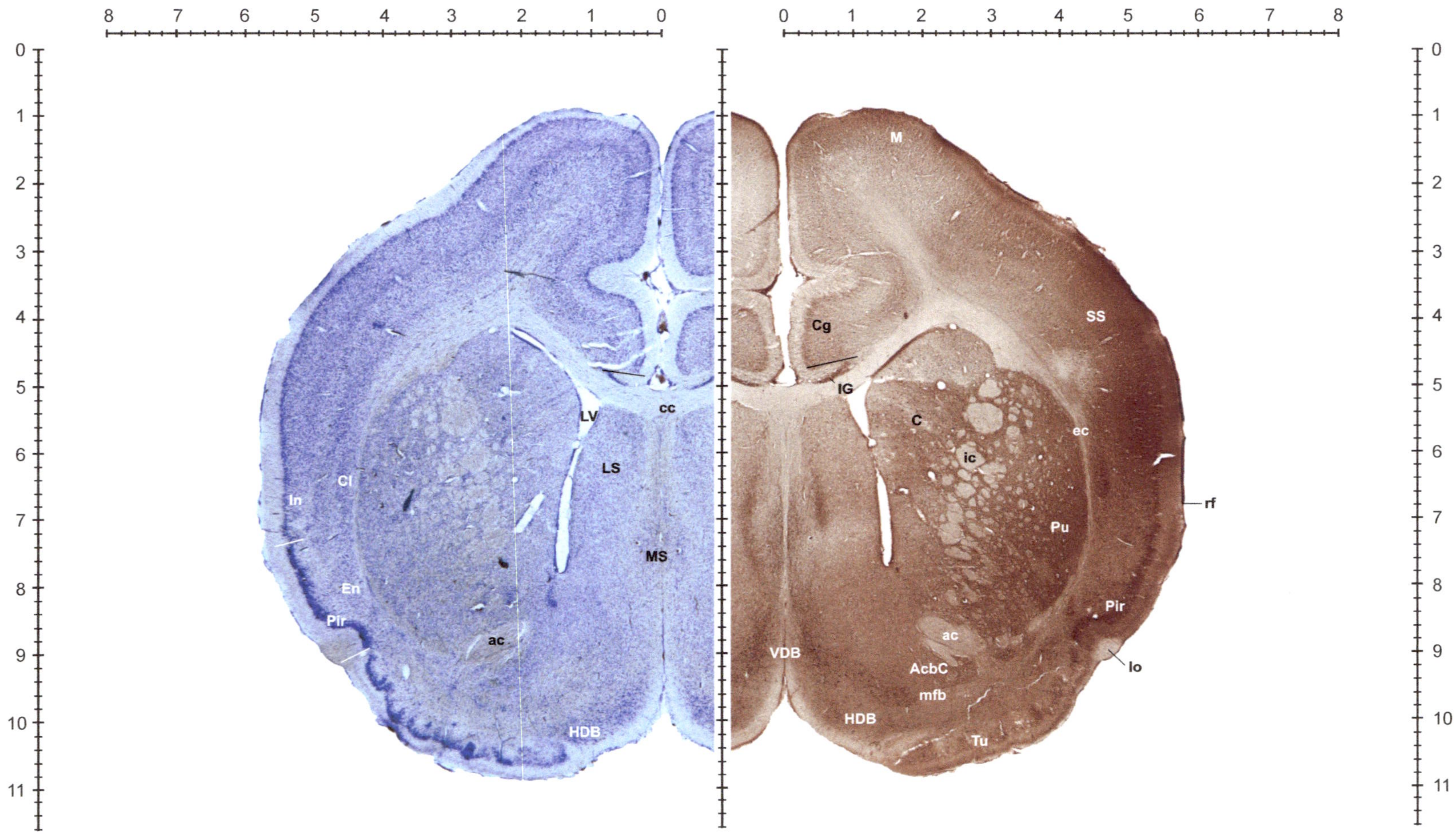


**ac** anterior commissure
**AcbC** accumbens nucleus, core
**C** caudate
**cc** corpus callosum
**Cg** cingulate cortex
**Cl** claustrum
**ec** external capsule
**En** endopiriform nucleus
**HDB** horizontal limb nu diagonal band
**ic** internal capsule
**IG** indusium griseum
**In** insular cortex
**lo** lateral olfactory tract
**LS** lateral septal nucleus
**LV** lateral ventricle
**M** motor cortex
**mfb** medial forebrain bundle
**MS** medial septal nucleus
**Pir** piriform cortex
**Pu** putamen
**rf** rhinal fissure
**SS** somatosensory cortex
**Tu** olfactory tubercle
**VDB** vertical limb nu diagonal band

## Plate 32 - Sinus: 8.50 mm

Interaural: 11.20 mm
Nissl section 76

Interaural: 11.15 mm
AChE section 76

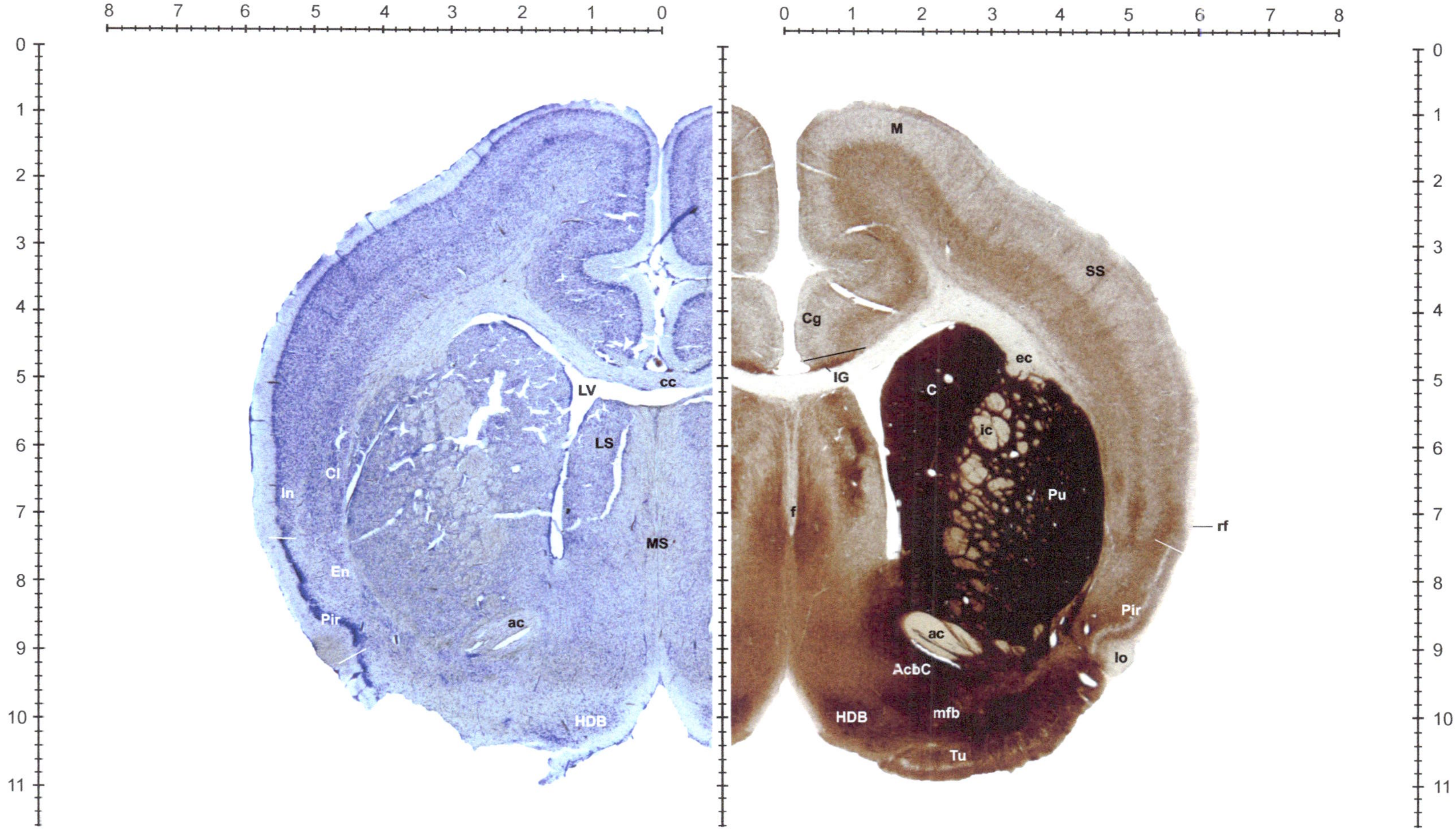


**ac** anterior commissure
**AcbC** accumbens nucleus, core
**C** caudate
**cc** corpus callosum
**Cg** cingulate cortex
**Cl** claustrum
**ec** external capsule
**En** endopiriform nucleus
**f** fornix
**HDB** horizontal limb nu diagonal band
**ic** internal capsule
**IG** indusium griseum
**In** insular cortex
**lo** lateral olfactory tract
**LS** lateral septal nucleus
**LV** lateral ventricle
**M** motor cortex
**mfb** medial forebrain bundle
**MS** medial septal nucleus
**Pir** piriform cortex
**Pu** putamen
**rf** rhinal fissure
**SS** somatosensory cortex
**Tu** olfactory tubercle

**Plate 33 - Sinus: 8.41 mm**

Interaural: 11.11 mm
Nissl section 77

Interaural: 10.99 mm
AChE section 78

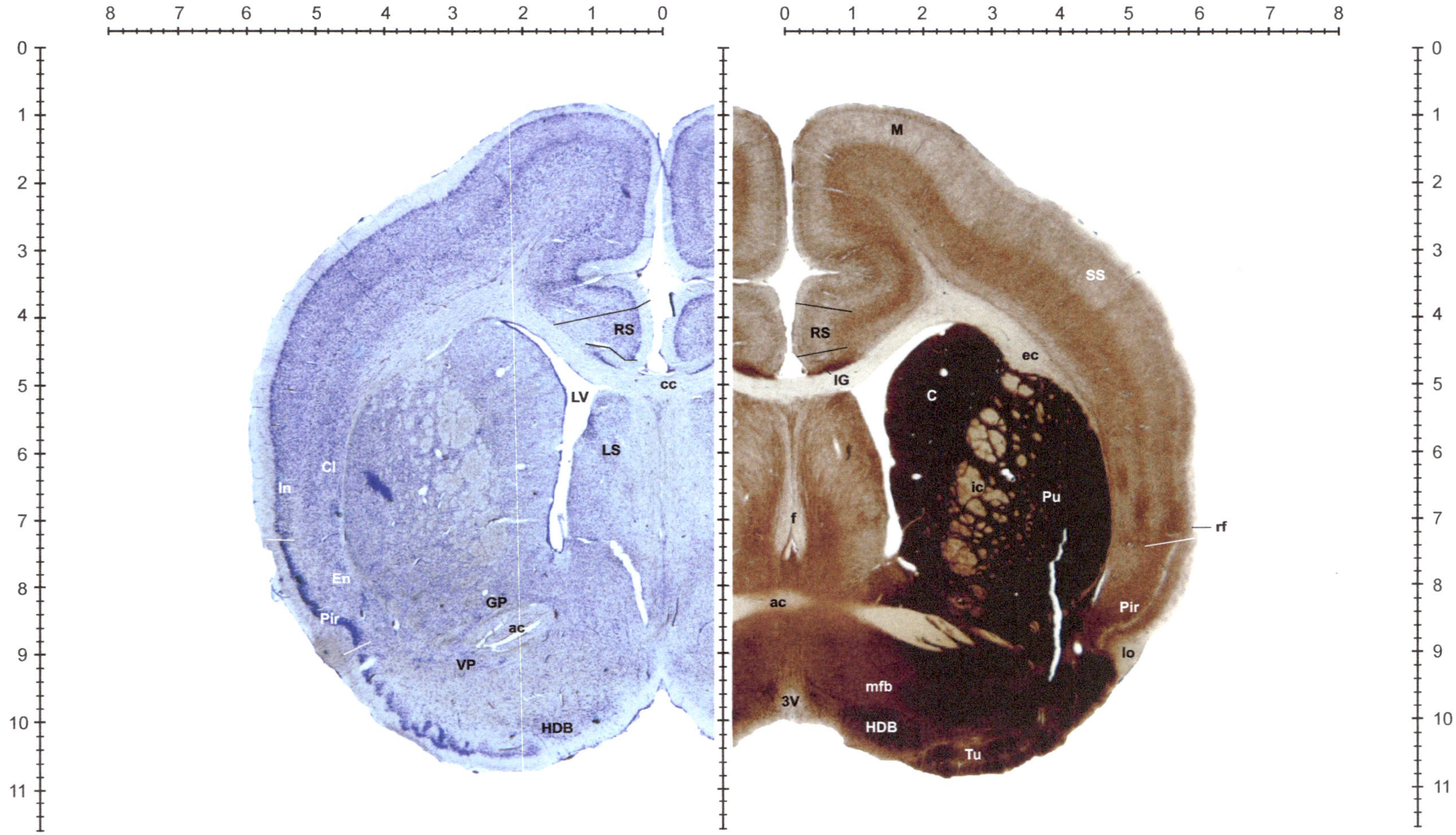


- **3V** 3rd ventricle
- **ac** anterior commissure
- **C** caudate
- **cc** corpus callosum
- **Cl** claustrum
- **ec** external capsule
- **En** endopiriform nucleus
- **f** fornix
- **GP** globus pallidus
- **HDB** horizontal limb nu diagonal band
- **ic** internal capsule
- **IG** indusium griseum
- **In** insular cortex
- **lo** lateral olfactory tract
- **LS** lateral septal nucleus
- **LV** lateral ventricle
- **M** motor cortex
- **mfb** medial forebrain bundle
- **Pir** piriform cortex
- **Pu** putamen
- **rf** rhinal fissure
- **RS** retrosplenial cortex
- **SS** somatosensory cortex
- **Tu** olfactory tubercle
- **VP** ventral pallidum

**Plate 34 - Sinus: 8.17 mm**

Interaural: 10.87 mm
Nissl section 80

nteraural: 10.74 mm
AChE section 81

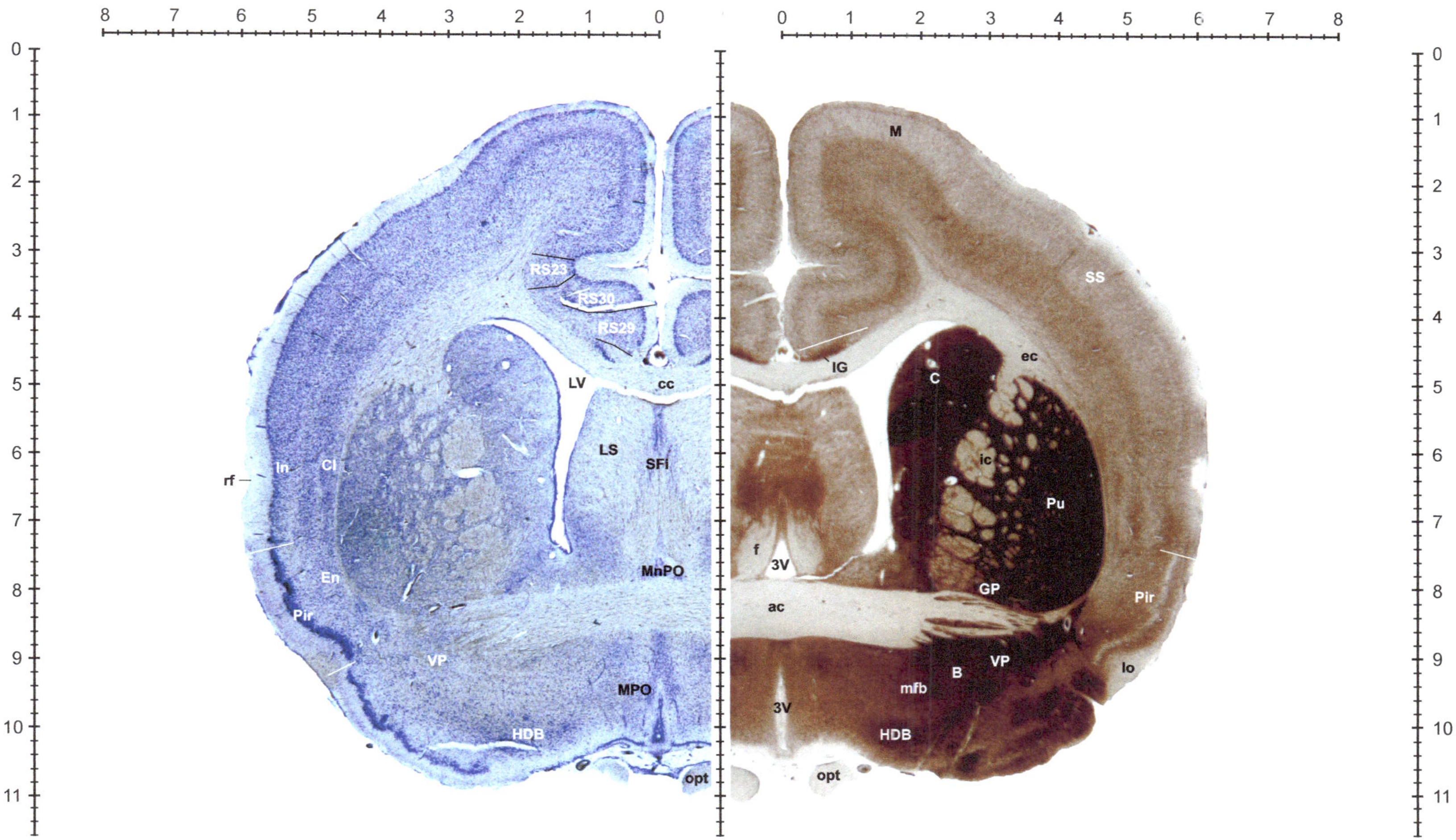


**3V** 3rd ventricle
**ac** anterior commissure
**B** basal nucleus (nucleus basalis)
**C** caudate
**cc** corpus callosum
**Cl** claustrum
**ec** external capsule
**En** endopiriform nucleus
**f** fornix
**GP** globus pallidus
**HDB** horizontal limb nu diagonal band
**ic** internal capsule
**IG** indusium griseum
**In** insular cortex
**lo** lateral olfactory tract
**LS** lateral septal nucleus
**LV** lateral ventricle
**M** motor cortex
**mfb** medial forebrain bundle
**MnPO** median preoptic nucleus
**MPO** medial preoptic nucleus
**opt** optic tract
**Pir** piriform cortex
**Pu** putamen
**rf** rhinal fissure
**RS23** retrosplenial ctx area 23
**RS29** retrosplenial ctx area 29
**RS30** retrosplenial ctx area 30
**SFi** septofimbrial nucleus
**SS** somatosensory cortex
**VP** ventral pallidum

**Plate 35 - Sinus: 7.92 mm**

Interaural: 10.62 mm
Nissl section 83

Interaural: 10.58 mm
Parv section 83

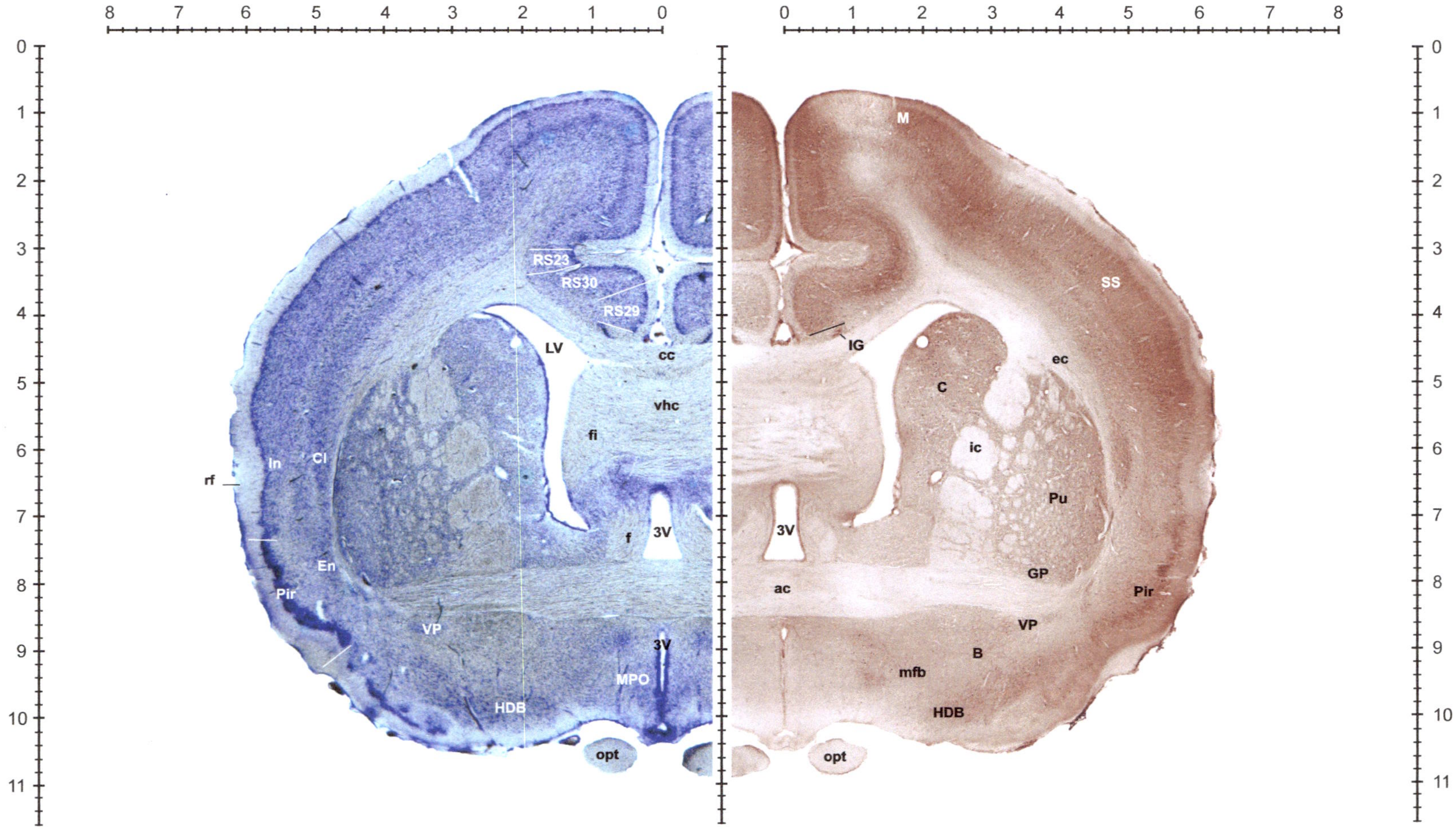


**3V** 3rd ventricle
**ac** anterior commissure
**B** basal nucleus (nucleus basalis)
**C** caudate
**cc** corpus callosum
**Cl** claustrum
**ec** external capsule
**En** endopiriform nucleus
**f** fornix
**fi** fimbria of the hippocampus
**GP** globus pallidus
**HDB** horizontal limb nu diagonal band
**ic** internal capsule
**IG** indusium griseum
**In** insular cortex
**LV** lateral ventricle
**M** motor cortex
**mfb** medial forebrain bundle
**MPO** medial preoptic nucleus
**opt** optic tract
**Pir** piriform cortex
**Pu** putamen
**rf** rhinal fissure
**RS23** retrosplenial ctx area 23
**RS29** retrosplenial ctx area 29
**RS30** retrosplenial ctx area 30
**SS** somatosensory cortex
**vhc** ventral hippocampal commissure
**VP** ventral pallidum

## Plate 36 - Sinus: 7.84 mm

Interaural: 10.54 mm
Nissl section 84

Interaural: 10.50 mm
AChE section 84

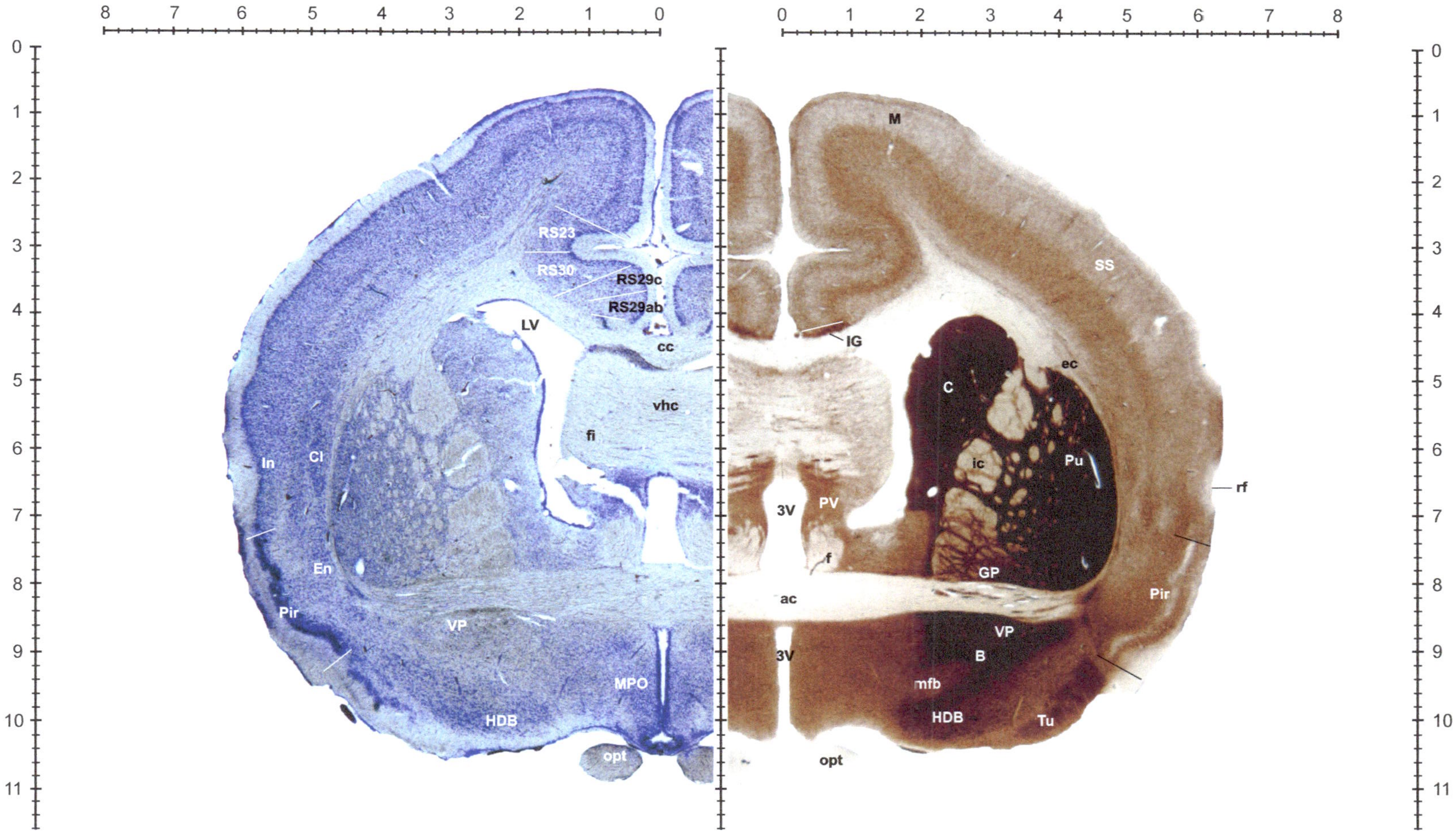


**3V** 3rd ventricle
**B** basal nucleus (nucleus basalis)
**C** caudate
**cc** corpus callosum
**Cl** claustrum
**ec** external capsule
**En** endopiriform nucleus
**f** fornix
**fi** fimbria of the hippocampus
**GP** globus pallidus
**HDB** horizontal limb nu diagonal band
**ic** internal capsule
**IG** indusium griseum
**In** insular cortex
**LV** lateral ventricle
**M** motor cortex
**mfb** medial forebrain bundle
**MPO** medial preoptic nucleus
**opt** optic tract
**Pir** piriform cortex
**Pu** putamen
**PV** paraventricular thalamic nucleus
**rf** rhinal fissure
**RS23** retrosplenial ctx area 23
**RS29ab** retrosplenial ctx area 29ab
**RS29c** retrosplenial ctx area 29c
**RS30** retrosplenial ctx area 30
**SS** somatosensory cortex
**Tu** olfactory tubercle
**vhc** ventral hippocampal commissure
**VP** ventral pallidum

**Plate 37 - Sinus: 7.68 mm**

Interaural: 10.38 mm
Nissl section 86

Interaural: 10.25 mm
AChE section 87

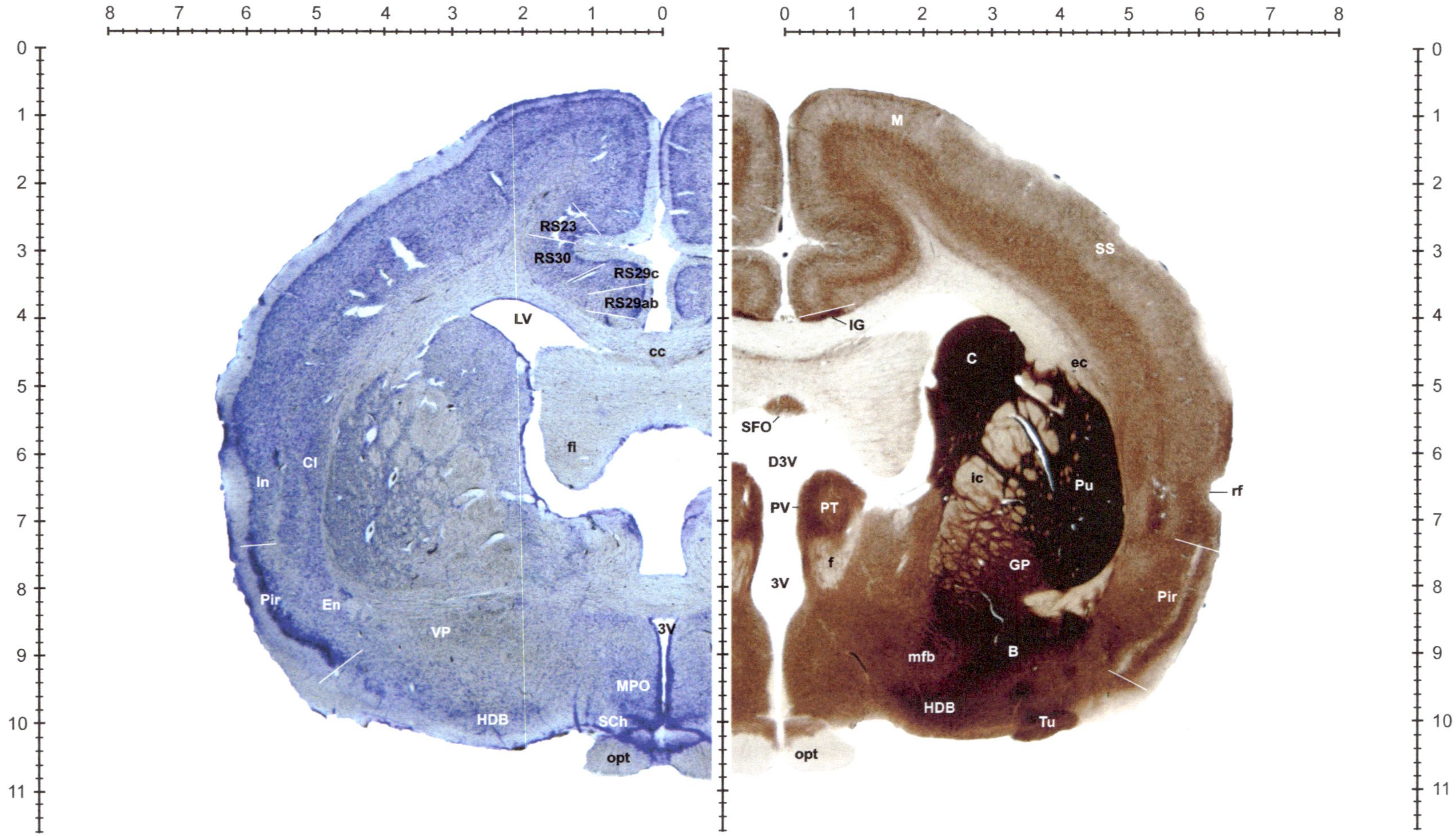


**3V** 3rd ventricle
**B** basal nucleus (nucleus basalis)
**C** caudate
**cc** corpus callosum
**Cl** claustrum
**D3V** dorsal 3rd ventricle
**En** endopiriform nucleus
**ec** external capsule
**f** fornix
**fi** fimbria of the hippocampus
**GP** globus pallidus
**HDB** horizontal limb nu diagonal band
**ic** internal capsule
**IG** indusium griseum
**In** insular cortex
**LV** lateral ventricle
**M** motor cortex
**mfb** medial forebrain bundle
**MPO** medial preoptic nucleus
**opt** optic tract
**Pir** piriform cortex
**PT** paratenial thalamic nucleus
**Pu** putamen
**PV** paraventricular thalamic nucleus
**rf** rhinal fissure
**RS23** retrosplenial ctx area 23
**RS29ab** retrosplenial ctx area 29ab
**RS29c** retrosplenial ctx area 29c
**RS30** retrosplenial ctx area 30
**SCh** suprachiasmatic nucleus
**SFO** subfornical organ
**SS** somatosensory cortex
**Tu** olfactory tubercle
**VP** ventral pallidum

## Plate 38 - Sinus: 7.51 mm

Interaural: 10.21 mm
Nissl section 88

Interaural: 10.09 mm
Parv section 89

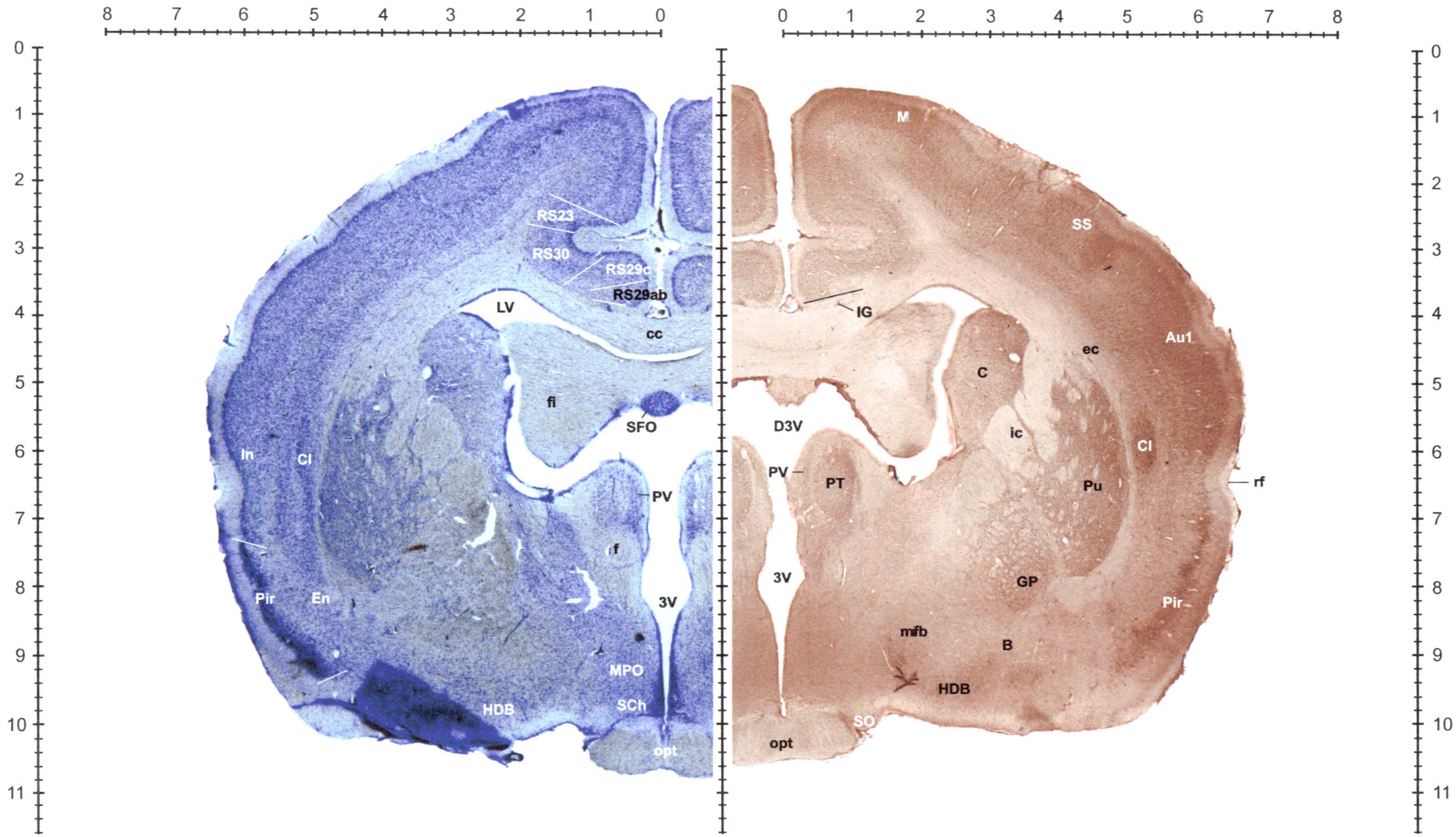


**3V** 3rd ventricle
**Au1** primary auditory cortex
**B** basal nucleus (nucleus basalis)
**C** caudate
**cc** corpus callosum
**Cl** claustrum
**D3V** dorsal 3rd ventricle
**ec** external capsule
**En** endopiriform nucleus
**f** fornix
**fi** fimbria of the hippocampus
**GP** globus pallidus
**HDB** horizontal limb nu diagonal band
**ic** internal capsule
**IG** indusium griseum
**In** insular cortex
**LV** lateral ventricle
**M** motor cortex
**mfb** medial forebrain bundle
**MPO** medial preoptic nucleus
**opt** optic tract
**Pir** piriform cortex
**PT** paratenial thalamic nucleus
**Pu** putamen
**PV** paraventricular thalamic nucleus
**rf** rhinal fissure
**RS23** retrosplenial ctx area 23
**RS29ab** retrosplenial ctx area 29ab
**RS29c** retrosplenial ctx area 29c
**RS30** retrosplenial ctx area 30
**SCh** suprachiasmatic nucleus
**SFO** subfornical organ
**SO** supraoptic nucleus
**SS** somatosensory cortex

## Plate 39 - Sinus: 7.35 mm

Interaural: 10.05 mm
Nissl section 90

Interaural: 10.01 mm
AChE section 90

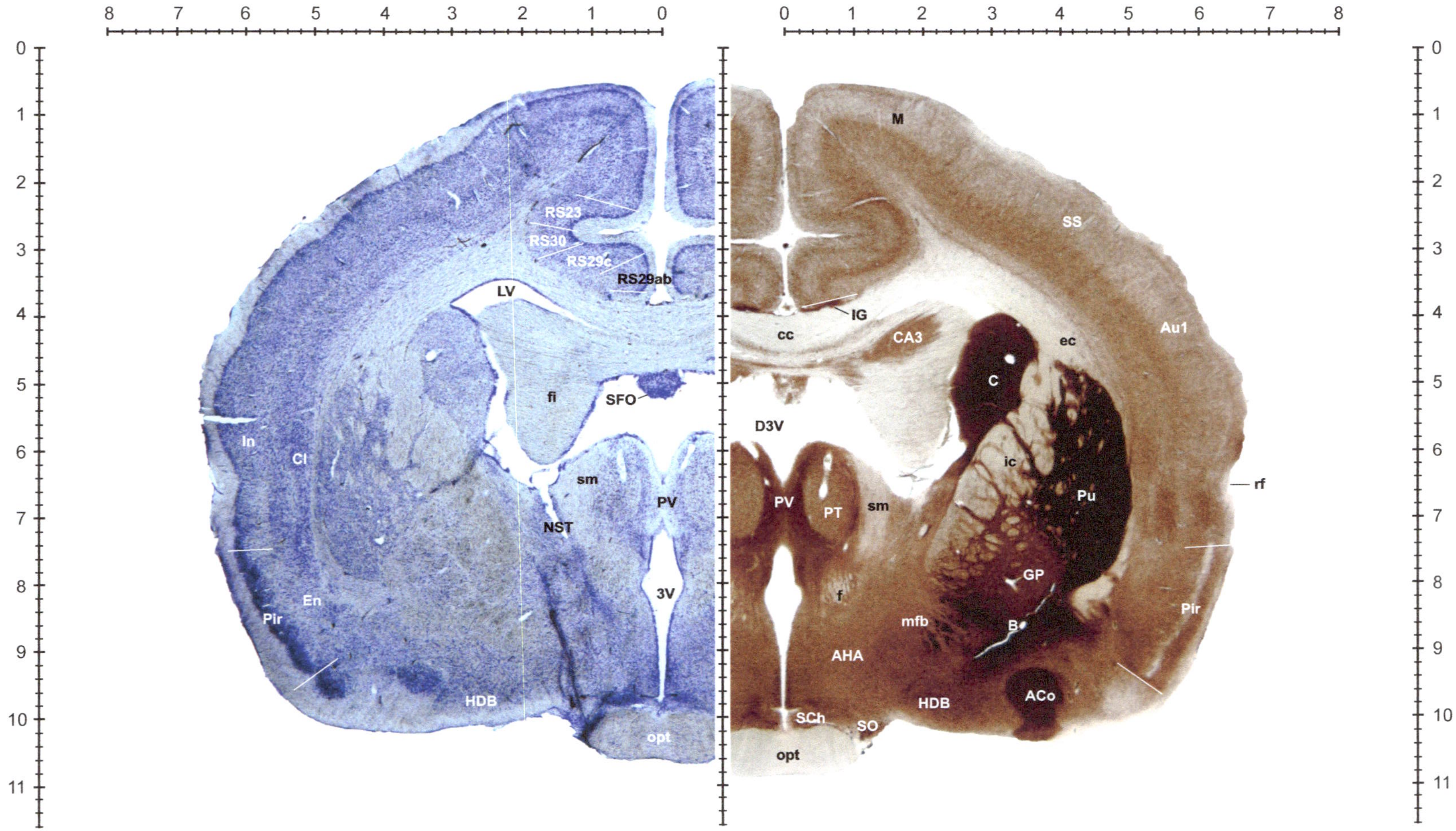


**3V** 3rd ventricle
**ACo** anterior cortical amygdaloid area
**AHA** anterior hypothal area, ant part
**Au1** primary auditory cortex
**B** basal nucleus (nucleus basalis)
**C** caudate
**CA3** field CA3 of the hippocampus
**cc** corpus callosum
**Cl** claustrum
**D3V** dorsal 3rd ventricle
**ec** external capsule
**En** endopiriform nucleus
**f** fornix
**fi** fimbria of the hippocampus
**GP** globus pallidus
**HDB** horizontal limb nu diagonal band
**ic** internal capsule
**IG** indusium griseum
**In** insular cortex
**LV** lateral ventricle
**M** motor cortex
**mfb** medial forebrain bundle
**NST** nu stria terminalis
**opt** optic tract
**Pir** piriform cortex
**PT** paratenial thalamic nucleus
**Pu** putamen
**PV** paraventricular thalamic nucleus
**rf** rhinal fissure
**RS23** retrosplenial ctx area 23
**RS29ab** retrosplenial ctx area 29ab
**RS29c** retrosplenial ctx area 29c
**RS30** retrosplenial ctx area 30
**SCh** suprachiasmatic nucleus
**SFO** subfornical organ
**sm** stria medullaris
**SO** supraoptic nucleus
**SS** somatosensory cortex

## Plate 40 - Sinus: 7.18 mm

Interaural: 9.88 mm
Nissl section 92

Interaural: 9.84 mm
Parv section 92

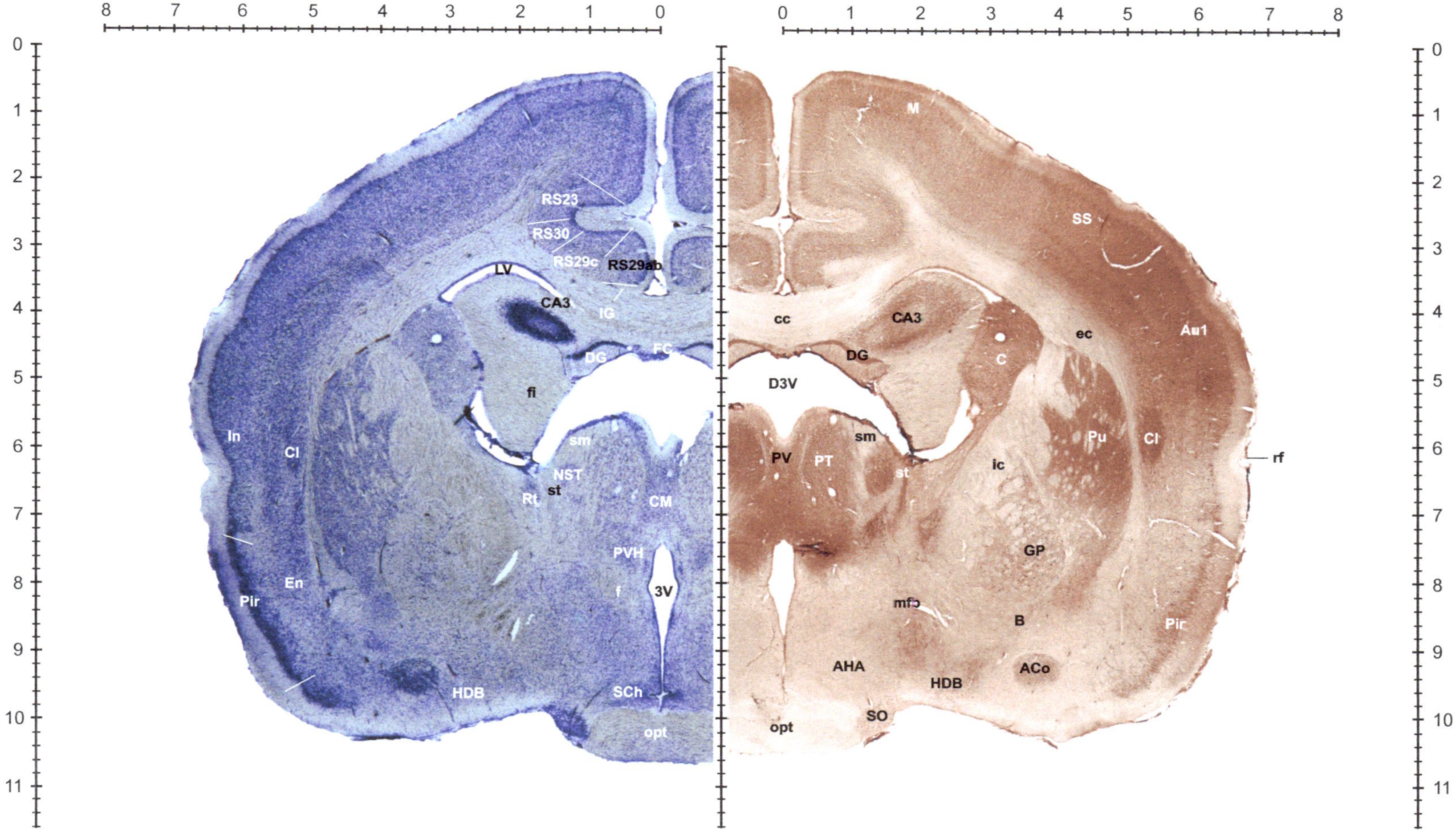


3V 3rd ventricle
ACo anterior cortical amygdaloid area
AHA anterior hypothal area, ant part
Au1 primary auditory cortex
B basal nucleus (nucleus basalis)
C caudate
CA3 field CA3 of the hippocampus
cc corpus callosum
Cl claustrum
CM central medial thalamic nucleus
D3V dorsal 3rd ventricle
DG dentate gyrus
ec external capsule
En endopiriform nucleus
f fornix
FC fasciola cinereum
fi fimbria of the hippocampus
GP globus pallidus
HDB horizontal limb nu diagonal band
ic internal capsule
IG indusium griseum
In insular cortex
LV lateral ventricle
M motor cortex
mfb medial forebrain bundle
NST nu stria terminalis
opt optic tract
Pir piriform cortex
PT paratenial thalamic nucleus
Pu putamen
PV paraventricular thalamic nucleus
PVH paraventricular hypothalamic nucleus
rf rhinal fissure
RS23 retrosplenial ctx area 23
RS29ab retrosplenial ctx area 29ab
RS29c retrosplenial ctx area 29c
RS30 retrosplenial ctx area 30
Rt reticular thalamic nucleus
SCh suprachiasmatic nucleus
sm stria medullaris
SO supraoptic nucleus
SS somatosensory cortex
st stria terminalis

**Plate 41 - Sinus: 7.02 mm**

Interaural: 9.72 mm

Nissl section 94

Interaural: 9.68 mm

AChE section 94

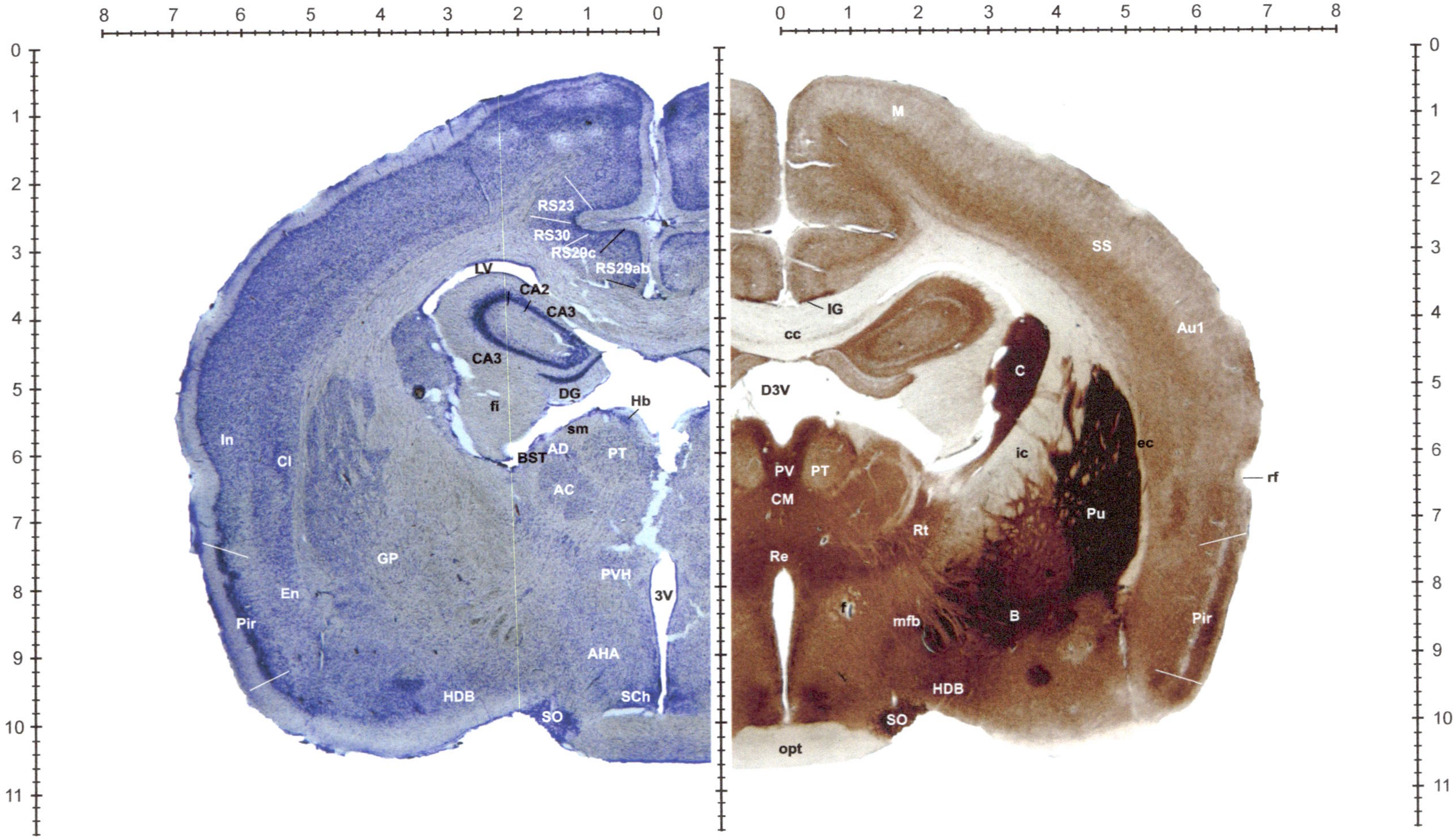


3V 3rd ventricle
AC anterior complex
AD anterodorsal thalamic nucleus
AHA anterior hypothal area, ant part
Au1 primary auditory cortex
B basal nucleus (nucleus basalis)
BST bed nu stria terminalis
C caudate
CA2 field CA2 of the hippocampus
CA3 field CA3 of the hippocampus
cc corpus callosum
Cl claustrum
CM central medial thalamic nucleus
D3V dorsal 3rd ventricle
DG dentate gyrus
ec external capsule
En endopiriform nucleus
f fornix
fi fimbria of the hippocampus
GP globus pallidus
Hb habenula
HDB horizontal limb nu diagonal band
ic internal capsule
IG indusium griseum
In insular cortex
LV lateral ventricle
M motor cortex
mfb medial forebrain bundle
opt optic tract
Pir piriform cortex
PT paratenial thalamic nucleus
Pu putamen
PV paraventricular thalamic nucleus
PVH paraventricular hypothalamic nucleus
rf rhinal fissure
RS23 retrosplenial ctx area 23
RS29ab retrosplenial ctx area 29ab
RS29c retrosplenial ctx area 29c
RS30 retrosplenial ctx area 30
Re reuniens thalamic nucleus
Rt reticular thalamic nucleus
SCh suprachiasmatic nucleus
sm stria medullaris
SO supraoptic nucleus
SS somatosensory cortex

## Plate 42 - Sinus: 6.77 mm

Interaural: 9.47 mm
Nissl section 97

Interaural: 9.43 mm
AChE section 97

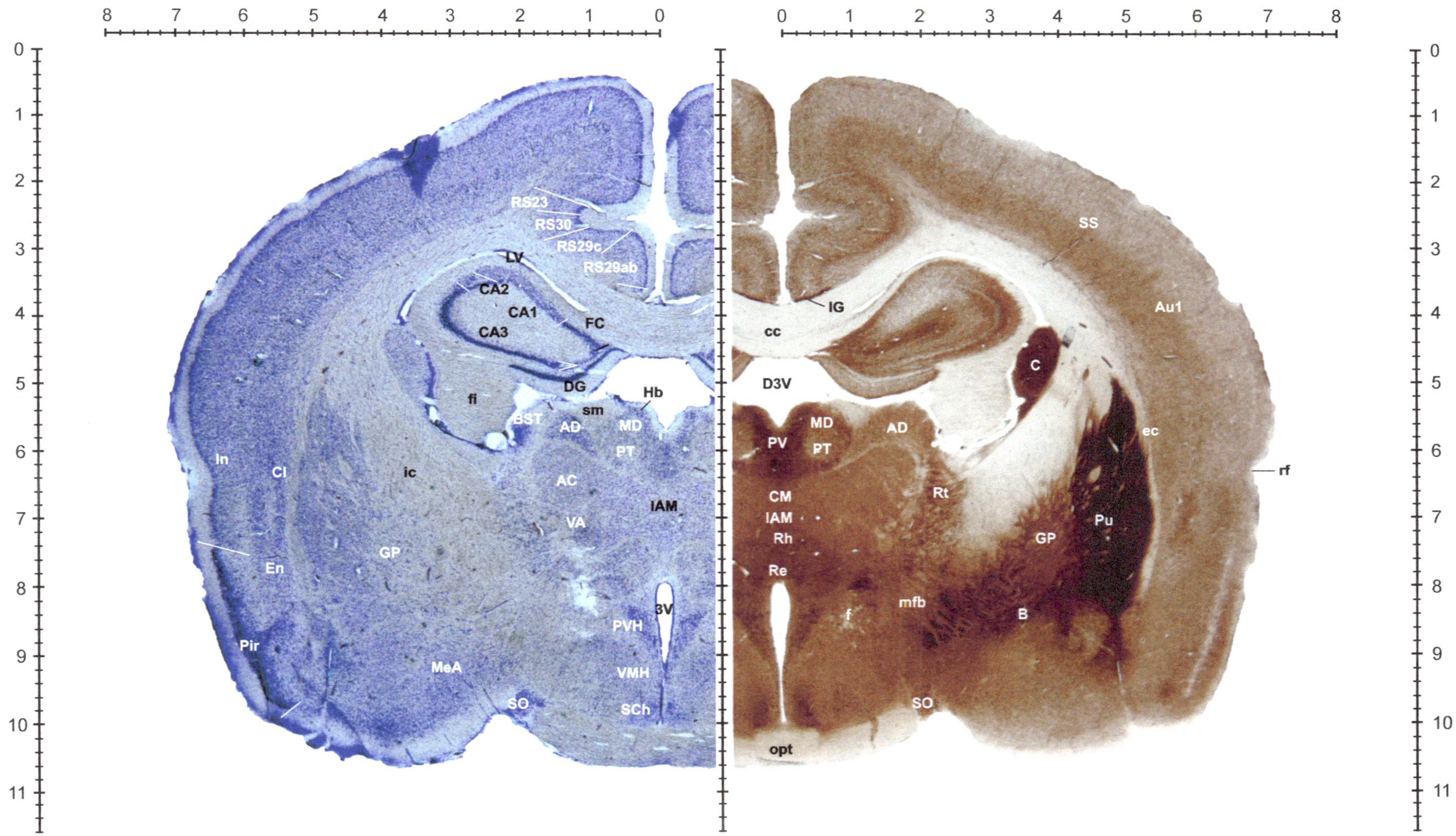


- **3V** 3rd ventricle
- **AC** anterior complex
- **AD** anterodorsal thalamic nucleus
- **Au1** primary auditory cortex
- **B** basal nucleus (nucleus basalis)
- **BST** bed nu stria terminalis
- **C** caudate
- **CA1** field CA1 of the hippocampus
- **CA2** field CA2 of the hippocampus
- **CA3** field CA3 of the hippocampus
- **cc** corpus callosum
- **Cl** claustrum
- **CM** central medial thalamic nucleus
- **D3V** dorsal 3rd ventricle
- **DG** dentate gyrus
- **ec** external capsule
- **En** endopiriform nucleus
- **f** fornix
- **FC** fasciola cinereum
- **fi** fimbria of the hippocampus
- **GP** globus pallidus
- **Hb** habenula
- **IAM** interanteromedial thalamic nucleus
- **ic** internal capsule
- **IG** indusium griseum
- **In** insular cortex
- **LV** lateral ventricle
- **MD** mediodorsal thalamic nucleus
- **MeA** medial amygdaloid nucleus, ant part
- **mfb** medial forebrain bundle
- **opt** optic tract
- **Pir** piriform cortex
- **PT** paratenial thalamic nucleus
- **Pu** putamen
- **PV** paraventricular thalamic nucleus
- **PVH** paraventricular hypothalamic nucleus
- **rf** rhinal fissure
- **RS23** retrosplenial ctx area 23
- **RS29ab** retrosplenial ctx area 29ab
- **RS29c** retrosplenial ctx area 29c
- **RS30** retrosplenial ctx area 30
- **Re** reuniens thalamic nucleus
- **Rh** rhomboid thalamic nucleus
- **Rt** reticular thalamic nucleus
- **SCh** suprachiasmatic nucleus
- **sm** stria medullaris
- **SO** supraoptic nucleus
- **SS** somatosensory cortex
- **VA** ventral anterior thalamic nucleus
- **VMH** ventromedial hypothalamic nucleus

## Plate 43 - Sinus: 6.69 mm

Interaural: 9.39 mm
Nissl section 98

Interaural: 9.35 mm
Parv section 98

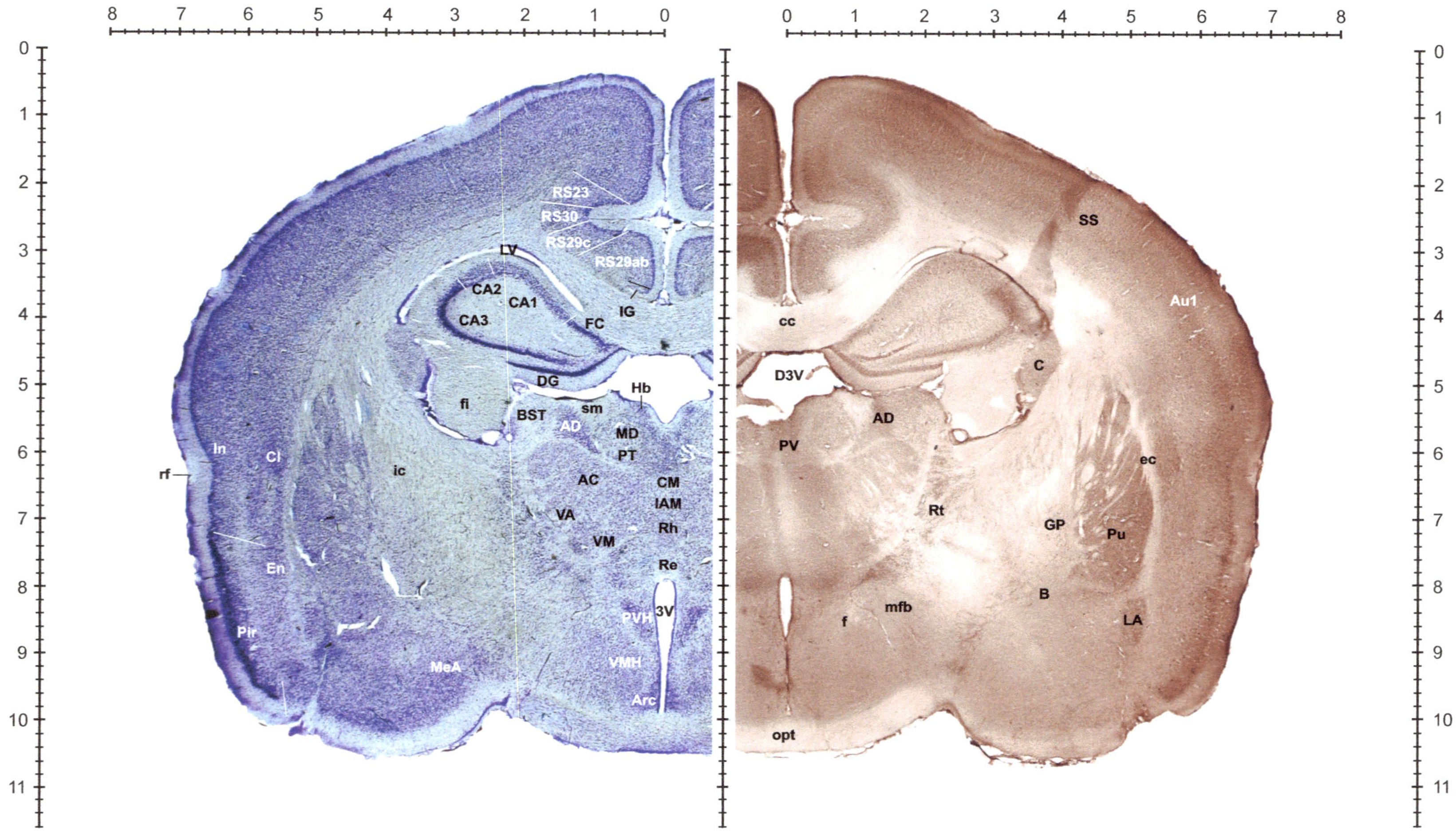


3V 3rd ventricle
AC anterior complex
AD anterodorsal thalamic nucleus
Arc arcuate hypothalamic nucleus
Au1 primary auditory cortex
B basal nucleus (nucleus basalis)
BST bed nu stria terminalis
C caudate
CA1 field CA1 of the hippocampus
CA2 field CA2 of the hippocampus
CA3 field CA3 of the hippocampus

cc corpus callosum
Cl claustrum
CM central medial thalamic nucleus
D3V dorsal 3rd ventricle
DG dentate gyrus
ec external capsule
En endopiriform nucleus
f fornix
FC fasciola cinereum
fi fimbria of the hippocampus
GP globus pallidus

Hb habenula
IAM interanteromedial thalamic nucleus
ic internal capsule
IG indusium griseum
In insular cortex
LA lateral amygdaloid nucleus
LV lateral ventricle
MD mediodorsal thalamic nucleus
MeA medial amygdaloid nucleus, ant part
mfb medial forebrain bundle
opt optic tract

Pir piriform cortex
PT paratenial thalamic nucleus
Pu putamen
PV paraventricular thalamic nucleus
PVH paraventricular hypothalamic nucleus
rf rhinal fissure
RS23 retrosplenial ctx area 23
RS29ab retrosplenial ctx area 29ab
RS29c retrosplenial ctx area 29c
RS30 retrosplenial ctx area 30
Re reuniens thalamic nucleus

Rh rhomboid thalamic nucleus
Rt reticular thalamic nucleus
sm stria medullaris
SS somatosensory cortex
VA ventral anterior thalamic nucleus
VM ventromedial thalamic nucleus
VMH ventromedial hypothalamic nucleus

**Plate 44 - Sinus: 6.61 mm**

Interaural: 9.31 mm
Nissl section 99

Interaural: 9.19 mm
AChE section 100

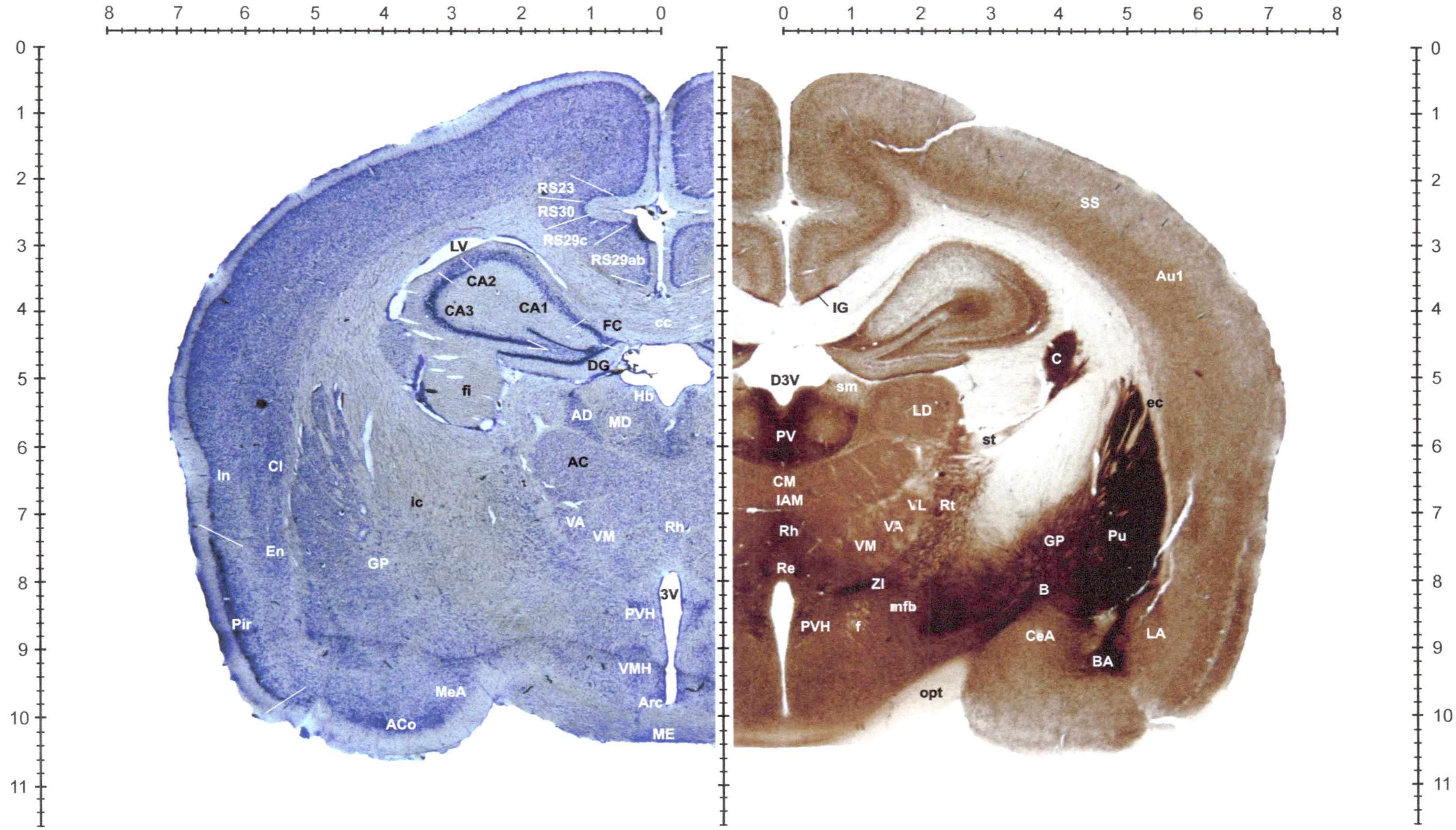


3V 3rd ventricle
AC anterior complex
ACo anterior cortical amygdaloid area
AD anterodorsal thalamic nucleus
Arc arcuate hypothalamic nucleus
Au1 primary auditory cortex
B basal nucleus (nucleus basalis)
BA basal amygdaloid nucleus
C caudate
CA1 field CA1 of the hippocampus
CA2 field CA2 of the hippocampus
CA3 field CA3 of the hippocampus
cc corpus callosum
CeA central amygdaloid nucleus
Cl claustrum
CM central medial thalamic nucleus
D3V dorsal 3rd ventricle
DG dentate gyrus
ec external capsule
En endopiriform nucleus
f fornix
FC fasciola cinereum
fi fimbria of the hippocampus
GP globus pallidus
Hb habenula
IAM interanteromedial thalamic nucleus
ic internal capsule
IG indusium griseum
In insular cortex
LA lateral amygdaloid nucleus
LD laterodorsal thalamic nucleus
LV lateral ventricle
MD mediodorsal thalamic nucleus
ME median eminence
MeA medial amygdaloid nucleus, ant part
mfb medial forebrain bundle
opt optic tract
Pir piriform cortex
Pu putamen
PV paraventricular thalamic nucleus
PVH paraventricular hypothalamic nucleus
Re reuniens thalamic nucleus
Rh rhomboid thalamic nucleus
RS23 retrosplenial ctx area 23
RS29ab retrosplenial ctx area 29ab
RS29c retrosplenial ctx area 29c
RS30 retrosplenial ctx area 30
Rt reticular thalamic nucleus
sm stria medullaris
SS somatosensory cortex
st stria terminalis
VA ventral anterior thalamic nucleus
VL ventral lateral thalamic nucleus
VM ventromedial thalamic nucleus
VMH ventromedial hypothalamic nucleus
ZI zona incerta

## Plate 45 - Sinus: 6.45 mm

Interaural: 9.15 mm
Nissl section 101

Interaural: 9.10 mm
Parv section 101

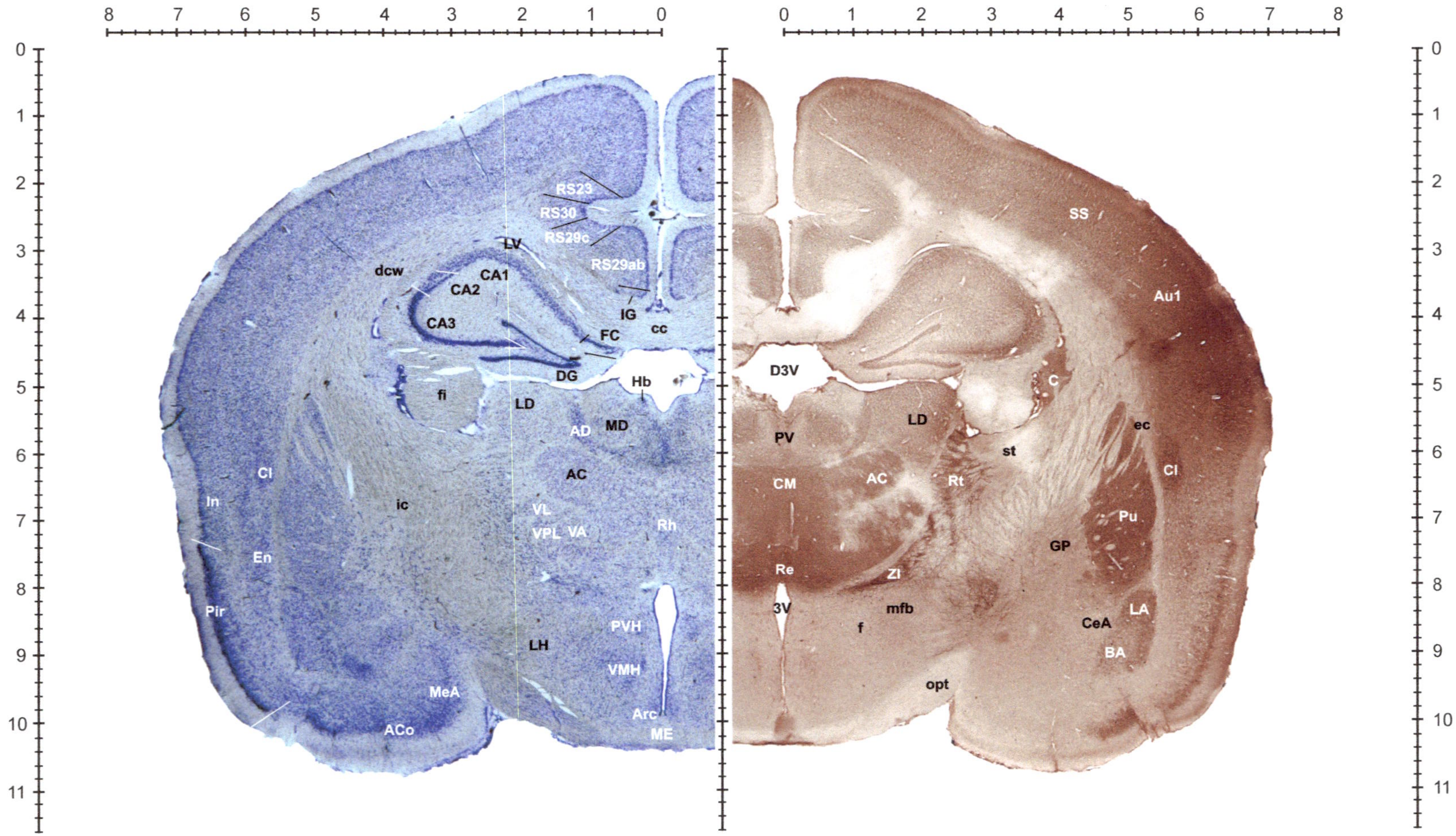


3V 3rd ventricle
AC anterior complex
ACo anterior cortical amygdaloid area
AD anterodorsal thalamic nucleus
Arc arcuate hypothalamic nucleus
Au1 primary auditory cortex
BA basal amygdaloid nucleus
C caudate
CA1 field CA1 of the hippocampus
CA2 field CA2 of the hippocampus
CA3 field CA3 of the hippocampus
cc corpus callosum
CeA central amygdaloid nucleus
Cl claustrum
CM central medial thalamic nucleus
D3V dorsal 3rd ventricle
dcw deep cerebral white matter
DG dentate gyrus
ec external capsule
En endopiriform nucleus
f fornix
FC fasciola cinereum
fi fimbria of the hippocampus
GP globus pallidus
Hb habenula
ic internal capsule
IG indusium griseum
In insular cortex
LA lateral amygdaloid nucleus
LD laterodorsal thalamic nucleus
LH lateral hypothalamic area
LV lateral ventricle
MD mediodorsal thalamic nucleus
ME median eminence
MeA medial amygdaloid nucleus, ant part
mfb medial forebrain bundle
opt optic tract
Pir piriform cortex
Pu putamen
PV paraventricular thalamic nucleus
PVH paraventricular hypothalamic nucleus
Re reuniens thalamic nucleus
Rh rhomboid thalamic nucleus
RS23 retrosplenial ctx area 23
RS29ab retrosplenial ctx area 29ab
RS29c retrosplenial ctx area 29c
RS30 retrosplenial ctx area 30
Rt reticular thalamic nucleus
SS somatosensory cortex
st stria terminalis
VA ventral anterior thalamic nucleus
VL ventral lateral thalamic nucleus
VMH ventromedial hypothalamic nucleus
VPL vent post lat thal nucleus
ZI zona incerta

**Plate 46 - Sinus: 6.36 mm**

Interaural: 9.06 mm
Nissl section 102

Interaural: 9.02 mm
AChE section 102

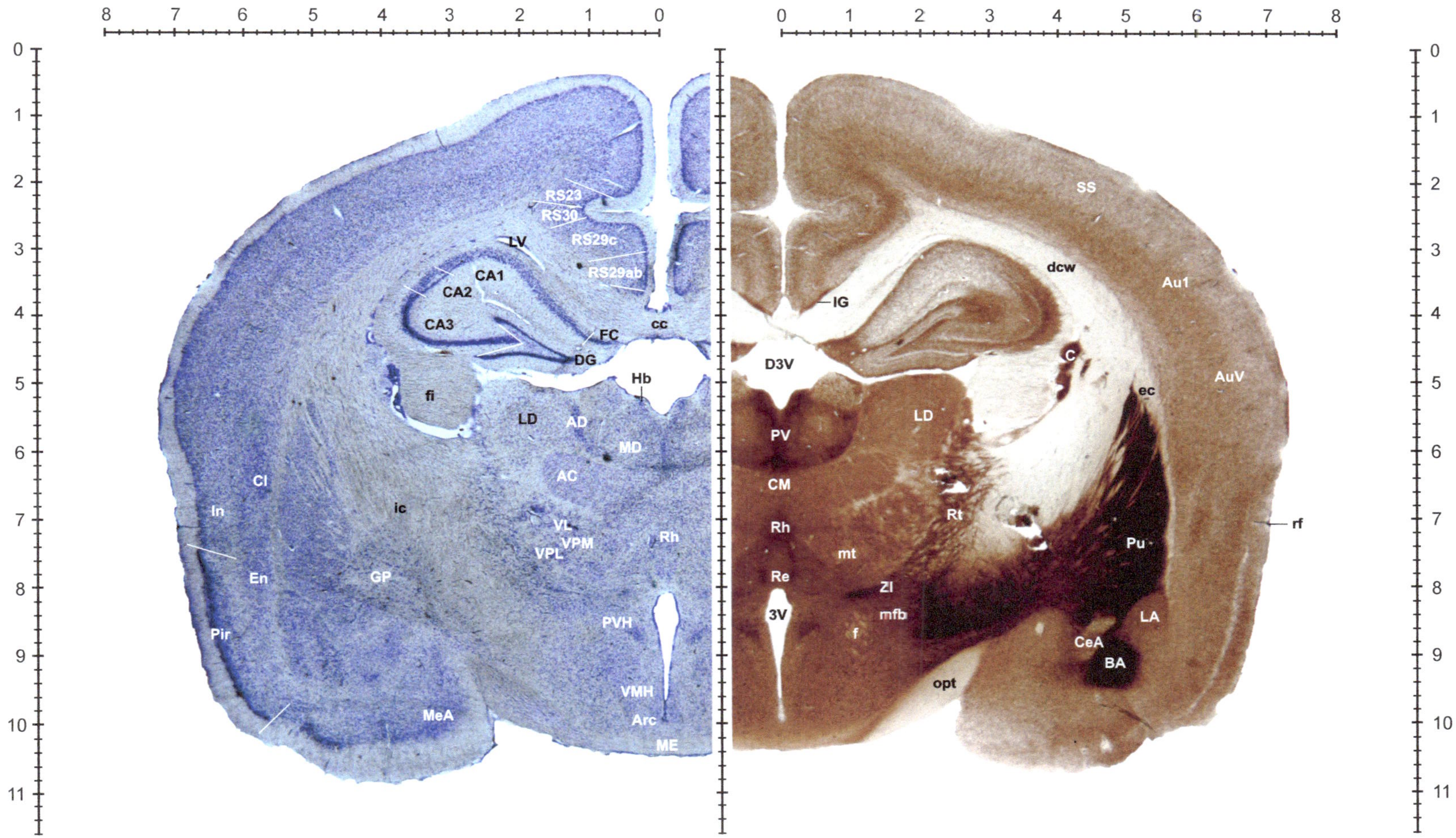


**3V** 3rd ventricle
**AC** anterior complex
**AD** anterodorsal thalamic nucleus
**Arc** arcuate hypothalamic nucleus
**Au1** primary auditory cortex
**AuV** secondary auditory ctx ventral area
**BA** basal amygdaloid nucleus
**C** caudate
**CA1** field CA1 of the hippocampus
**CA2** field CA2 of the hippocampus
**CA3** field CA3 of the hippocampus
**cc** corpus callosum
**CeA** central amygdaloid nucleus
**Cl** claustrum
**CM** central medial thalamic nucleus
**D3V** dorsal 3rd ventricle
**dcw** deep cerebral white matter
**DG** dentate gyrus
**ec** external capsule
**En** endopiriform nucleus
**f** fornix
**FC** fasciola cinereum
**fi** fimbria of the hippocampus
**GP** globus pallidus
**Hb** habenula
**ic** internal capsule
**IG** indusium griseum
**In** insular cortex
**LA** lateral amygdaloid nucleus
**LD** laterodorsal thalamic nucleus
**LV** lateral ventricle
**MD** mediodorsal thalamic nucleus
**ME** median eminence
**MeA** medial amygdaloid nucleus, ant part
**mfb** medial forebrain bundle
**mt** mammillothalamic tract
**opt** optic tract
**Pir** piriform cortex
**Pu** putamen
**PV** paraventricular thalamic nucleus
**PVH** paraventricular hypothalamic nucleus
**Re** reuniens thalamic nucleus
**rf** rhinal fissure
**Rh** rhomboid thalamic nucleus
**RS23** retrosplenial ctx area 23
**RS29ab** retrosplenial ctx area 29ab
**RS29c** retrosplenial ctx area 29c
**RS30** retrosplenial ctx area 30
**Rt** reticular thalamic nucleus
**SS** somatosensory cortex
**VL** ventral lateral thalamic nucleus
**VMH** ventromedial hypothalamic nucleus
**VPL** vent post lat thal nucleus
**VPM** vent post med thal nucleus
**ZI** zona incerta

## Plate 47 - Sinus: 6.28 mm

Interaural: 8.98 mm
Nissl section 103

Interaural: 8.94 mm
AChE section 103

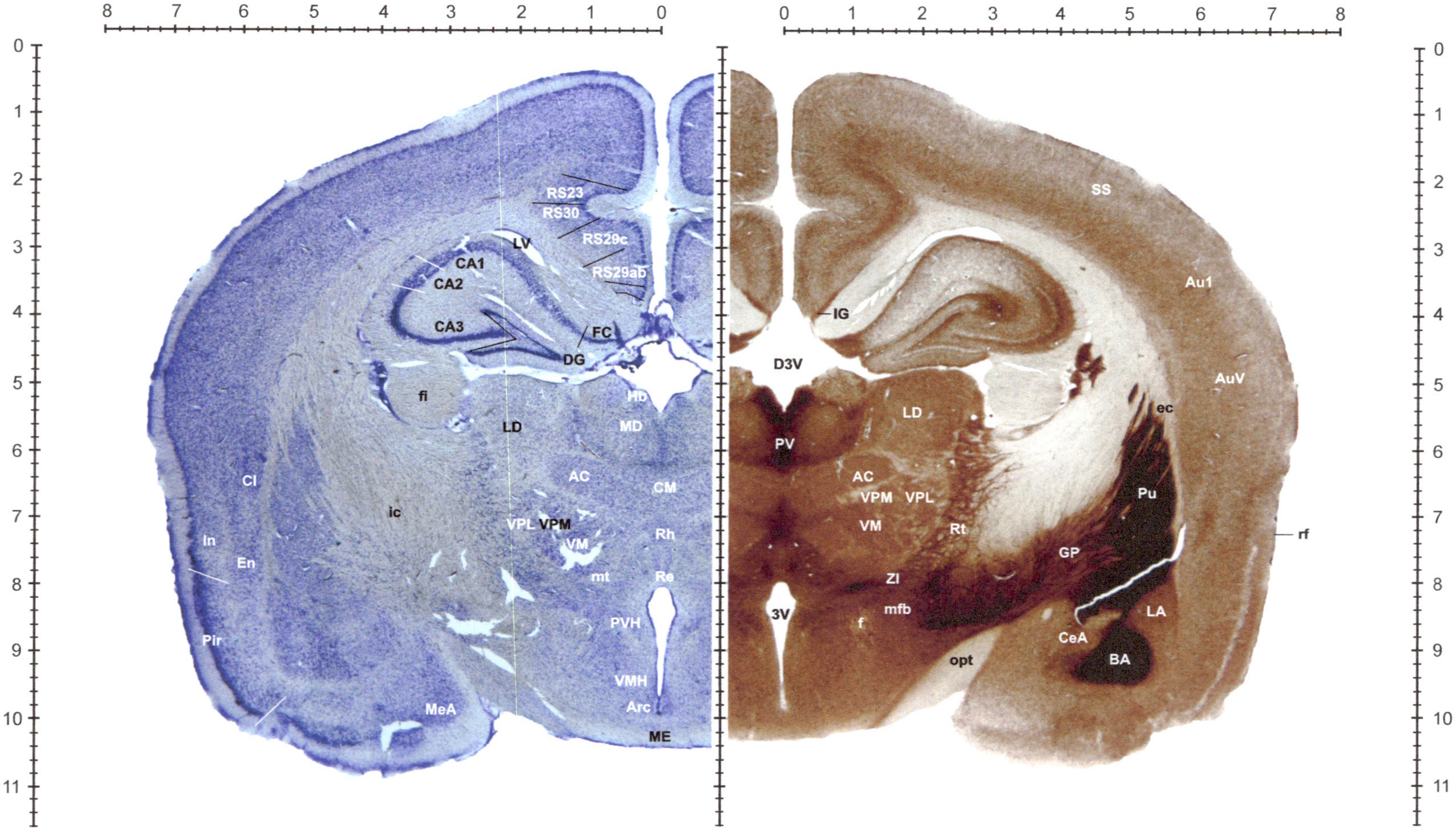


**3V** 3rd ventricle
**AC** anterior complex
**Arc** arcuate hypothalamic nucleus
**Au1** primary auditory cortex
**AuV** secondary auditory ctx ventral area
**BA** basal amygdaloid nucleus
**CA1** field CA1 of the hippocampus
**CA2** field CA2 of the hippocampus
**CA3** field CA3 of the hippocampus
**CeA** central amygdaloid nucleus
**Cl** claustrum
**CM** central medial thalamic nucleus
**D3V** dorsal 3rd ventricle
**DG** dentate gyrus
**ec** external capsule
**En** endopiriform nucleus
**f** fornix
**FC** fasciola cinereum
**fi** fimbria of the hippocampus
**GP** globus pallidus
**Hb** habenula
**ic** internal capsule
**IG** indusium griseum
**In** insular cortex
**LA** lateral amygdaloid nucleus
**LD** laterodorsal thalamic nucleus
**LV** lateral ventricle
**MD** mediodorsal thalamic nucleus
**ME** median eminence
**MeA** medial amygdaloid nucleus, ant part
**mfb** medial forebrain bundle
**mt** mammillothalamic tract
**opt** optic tract
**Pir** piriform cortex
**Pu** putamen
**PV** paraventricular thalamic nucleus
**PVH** paraventricular hypothalamic nucleus
**Re** reuniens thalamic nucleus
**rf** rhinal fissure
**Rh** rhomboid thalamic nucleus
**RS23** retrosplenial ctx area 23
**RS29ab** retrosplenial ctx area 29ab
**RS29c** retrosplenial ctx area 29c
**RS30** retrosplenial ctx area 30
**Rt** reticular thalamic nucleus
**SS** somatosensory cortex
**VM** ventromedial thalamic nucleus
**VMH** ventromedial hypothalamic nucleus
**VPL** vent post lat thal nucleus
**VPM** vent post med thal nucleus
**ZI** zona incerta

## Plate 48 - Sinus: 6.20 mm

Interaural: 8.90 mm
Nissl section 104

Interaural: 8.86 mm
Parv section 104

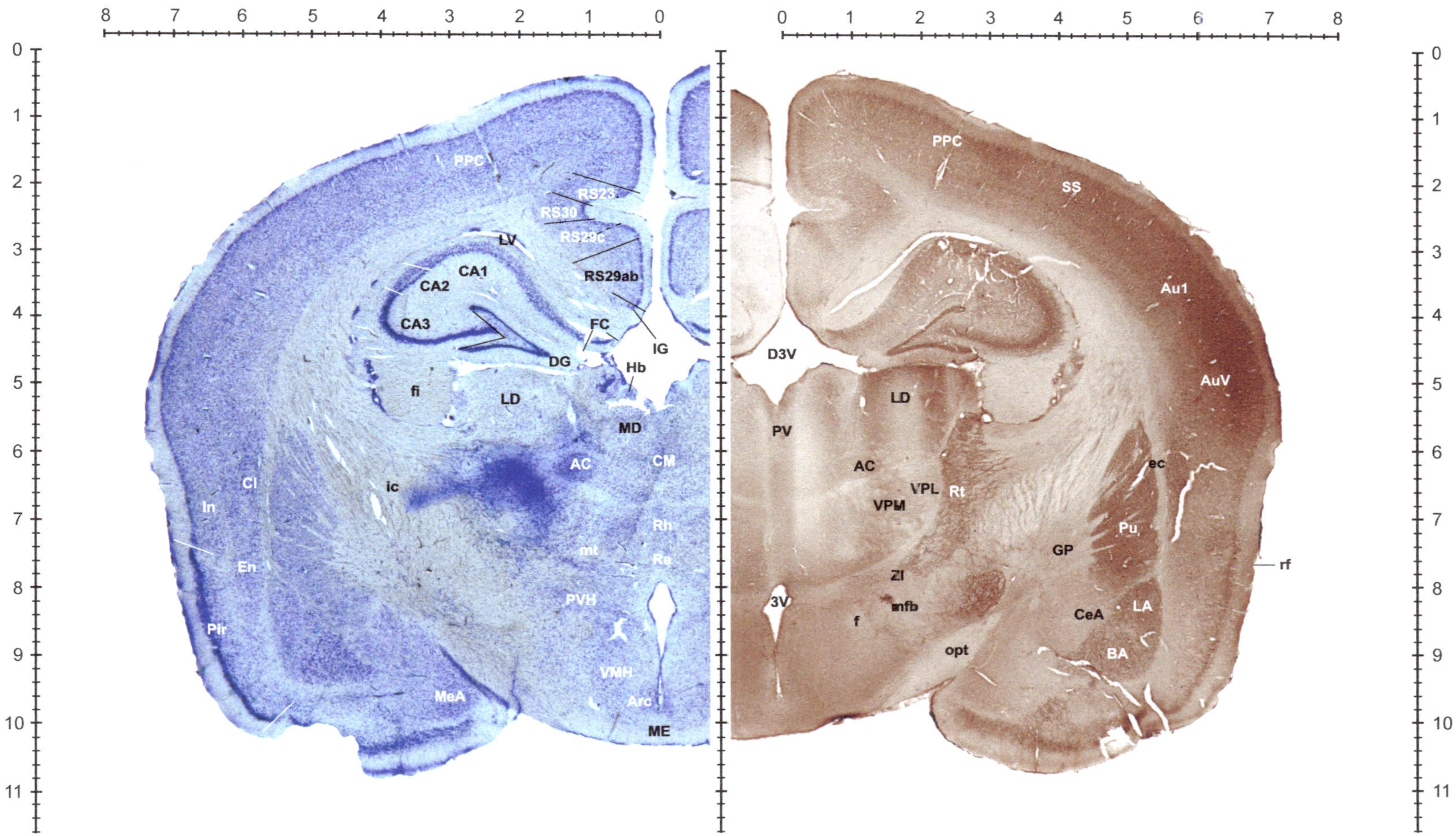


**3V** 3rd ventricle
**AC** anterior complex
**Arc** arcuate hypothalamic nucleus
**Au1** primary auditory cortex
**AuV** secondary auditory ctx ventral area
**BA** basal amygdaloid nucleus
**CA1** field CA1 of the hippocampus
**CA2** field CA2 of the hippocampus
**CA3** field CA3 of the hippocampus
**CeA** central amygdaloid nucleus
**Cl** claustrum
**CM** central medial thalamic nucleus
**D3V** dorsal 3rd ventricle
**DG** dentate gyrus
**ec** external capsule
**En** endopiriform nucleus
**f** fornix
**FC** fasciola cinereum
**fi** fimbria of the hippocampus
**GP** globus pallidus
**Hb** habenula
**ic** internal capsule
**IG** indusium griseum
**In** insular cortex
**LA** lateral amygdaloid nucleus
**LD** laterodorsal thalamic nucleus
**LV** lateral ventricle
**MD** mediodorsal thalamic nucleus
**ME** median eminence
**MeA** medial amygdaloid nucleus, ant part
**mfb** medial forebrain bundle
**mt** mammillothalamic tract
**opt** optic tract
**Pir** piriform cortex
**PPC** posterior parietal cortex
**Pu** putamen
**PV** paraventricular thalamic nucleus
**PVH** paraventricular hypothalamic nucleus
**Re** reuniens thalamic nucleus
**rf** rhinal fissure
**Rh** rhomboid thalamic nucleus
**RS23** retrosplenial ctx area 23
**RS29ab** retrosplenial ctx area 29ab
**RS29c** retrosplenial ctx area 29c
**RS30** retrosplenial ctx area 30
**Rt** reticular thalamic nucleus
**SS** somatosensory cortex
**VMH** ventromedial hypothalamic nucleus
**VPL** vent post lat thal nucleus
**VPM** vent post med thal nucleus
**ZI** zona incerta

## Plate 49 - Sinus: 6.04 mm

Interaural: 8.74 mm
Nissl section 106

Interaural: 8.69 mm
AChE section 106

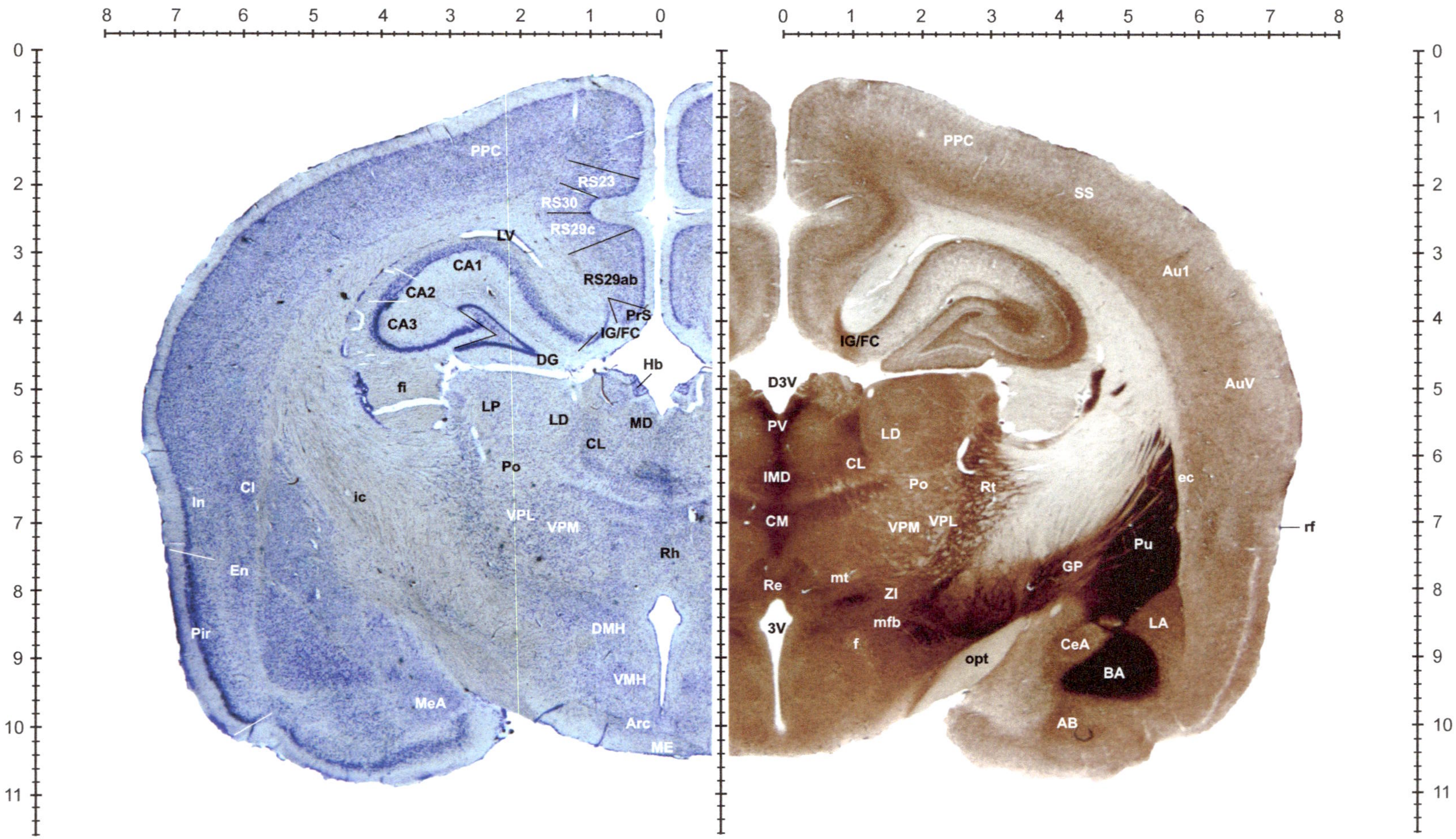


**3V** 3rd ventricle
**AB** accessory basal nucleus amygdala
**Arc** arcuate hypothalamic nucleus
**Au1** primary auditory cortex
**AuV** secondary auditory ctx ventral area
**BA** basal amygdaloid nucleus
**CA1** field CA1 of the hippocampus
**CA2** field CA2 of the hippocampus
**CA3** field CA3 of the hippocampus
**CeA** central amygdaloid nucleus
**CL** centrolateral thalamic nucleus
**Cl** claustrum
**CM** central medial thalamic nucleus
**D3V** dorsal 3rd ventricle
**DG** dentate gyrus
**DMH** dorsomedial hypothalamic nucleus
**ec** external capsule
**En** endopiriform nucleus
**f** fornix
**FC** fasciola cinereum
**fi** fimbria of the hippocampus
**GP** globus pallidus
**Hb** habenula
**ic** internal capsule
**IG** indusium griseum
**In** insular cortex
**IMD** intermediodorsal thalamic nucleus
**LA** lateral amygdaloid nucleus
**LD** laterodorsal thalamic nucleus
**LP** lateral posterior thalamic nucleus
**LV** lateral ventricle
**MD** mediodorsal thalamic nucleus
**ME** median eminence
**MeA** medial amygdaloid nucleus, ant part
**mfb** medial forebrain bundle
**mt** mammillothalamic tract
**opt** optic tract
**Pir** piriform cortex
**Po** posterior thalamic nuclear group
**PPC** posterior parietal cortex
**PrS** presubiculum
**Pu** putamen
**PV** paraventricular thalamic nucleus
**Re** reuniens thalamic nucleus
**rf** rhinal fissure
**Rh** rhomboid thalamic nucleus
**RS23** retrosplenial ctx area 23
**RS29ab** retrosplenial ctx area 29ab
**RS29c** retrosplenial ctx area 29c
**RS30** retrosplenial ctx area 30
**Rt** reticular thalamic nucleus
**SS** somatosensory cortex
**VMH** ventromedial hypothalamic nucleus
**VPL** vent post lat thal nucleus
**VPM** vent post med thal nucleus
**ZI** zona incerta

## Plate 50 - Sinus: 5.95 mm

Interaural: 8.65 mm
Nissl section 107

Interaural: 8.53 mm
AChE section 108

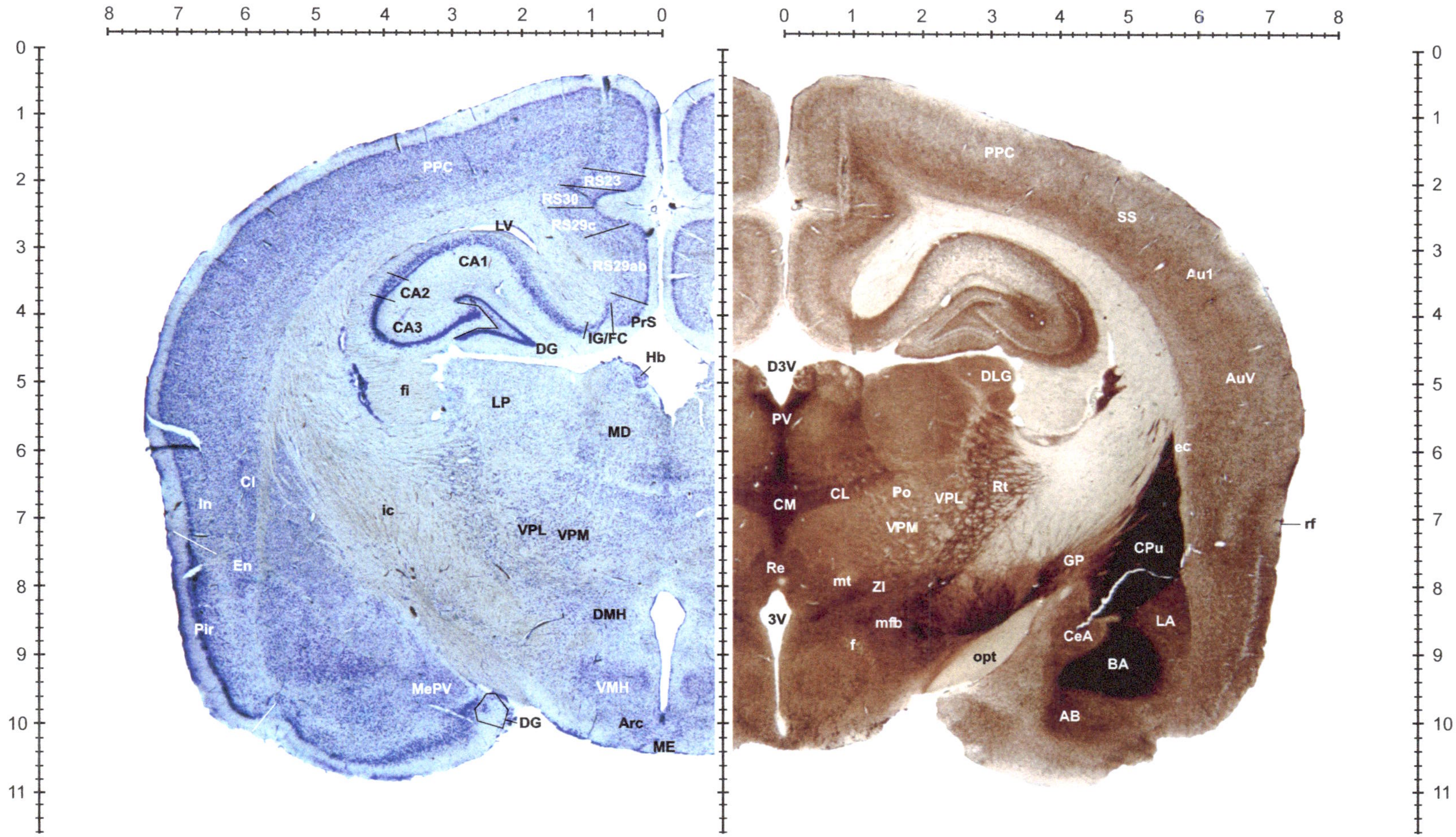


**3V** 3rd ventricle
**AB** accessory basal nucleus amygdala
**Arc** arcuate hypothalamic nucleus
**Au1** primary auditory cortex
**AuV** secondary auditory ctx ventral area
**BA** basal amygdaloid nucleus
**CA1** field CA1 of the hippocampus
**CA2** field CA2 of the hippocampus
**CA3** field CA3 of the hippocampus
**CeA** central amygdaloid nucleus
**CL** centrolateral thalamic nucleus
**Cl** claustrum
**CM** central medial thalamic nucleus
**CPu** caudate putamen (striatum)
**D3V** dorsal 3rd ventricle
**DG** dentate gyrus
**DLG** dorsal lateral geniculate nucleus
**DMH** dorsomedial hypothalamic nucleus
**ec** external capsule
**En** endopiriform nucleus
**f** fornix
**FC** fasciola cinereum
**fi** fimbria of the hippocampus
**GP** globus pallidus
**Hb** habenula
**ic** internal capsule
**IG** indusium griseum
**In** insular cortex
**LA** lateral amygdaloid nucleus
**LP** lateral posterior thalamic nucleus
**LV** lateral ventricle
**MD** mediodorsal thalamic nucleus
**ME** median eminence
**MeFV** med amy nucleus, post vent part
**mfb** medial forebrain bundle
**mt** mammillothalamic tract
**opt** optic tract
**Pir** piriform cortex
**Po** posterior thalamic nuclear group
**PPC** posterior parietal cortex
**PrS** presubiculum
**PV** paraventricular thalamic nucleus
**Re** reuniens thalamic nucleus
**rf** rhinal fissure
**RS23** retrosplenial ctx area 23
**RS29ab** retrosplenial ctx area 29ab
**RS29c** retrosplenial ctx area 29c
**RS30** retrosplenial ctx area 30
**Rt** reticular thalamic nucleus
**SS** somatosensory cortex
**VMH** ventromedial hypothalamic nucleus
**VPL** vent post lat thal nucleus
**VPM** vent post med thal nucleus
**ZI** zona incerta

## Plate 51 - Sinus: 5.79 mm

Interaural: 8.49 mm
Nissl section 109

Interaural: 8.45 mm
AChE section 109

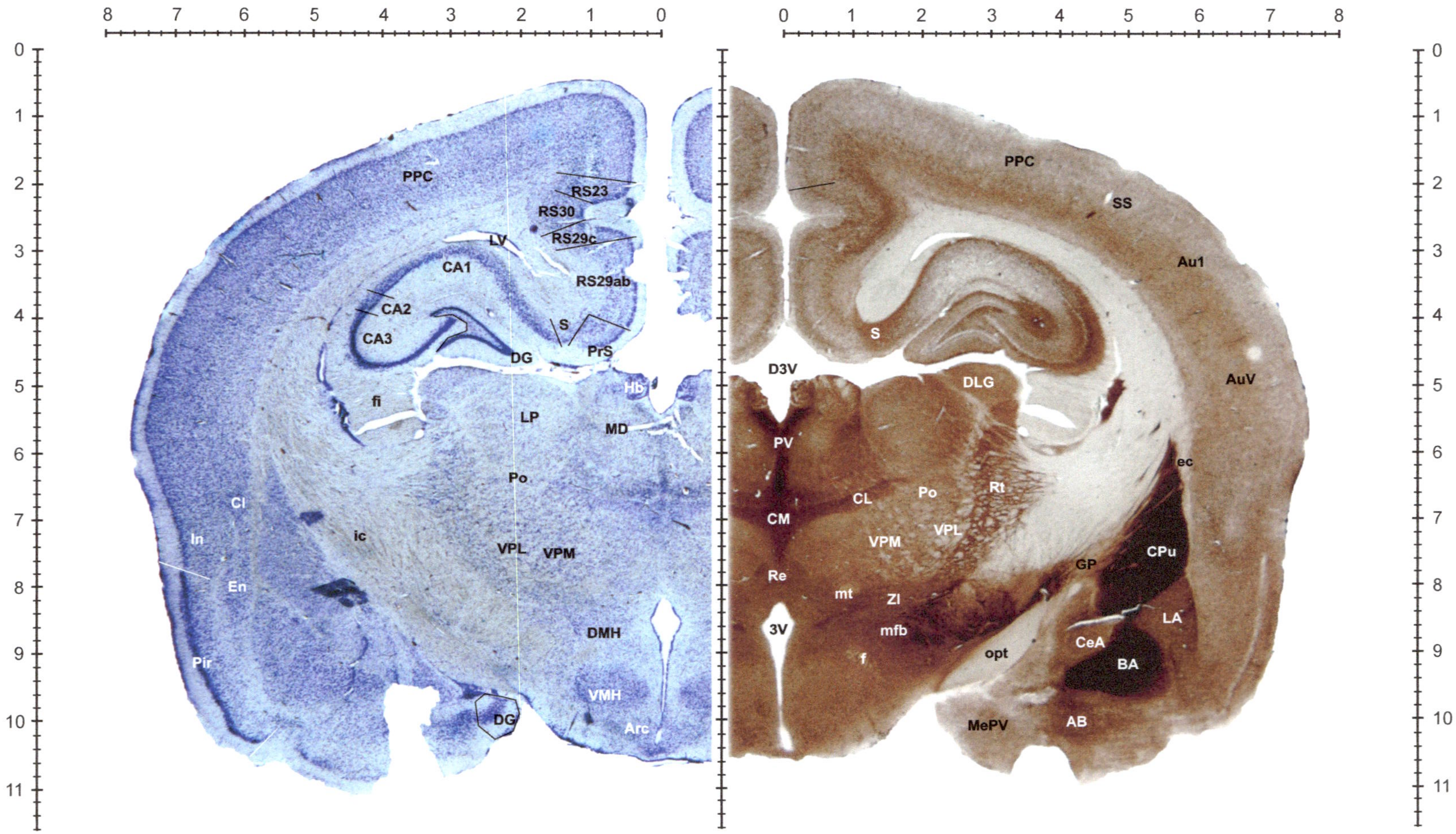


3V 3rd ventricle
AB accessory basal nucleus amygdala
Arc arcuate hypothalamic nucleus
Au1 primary auditory cortex
AuV secondary auditory ctx ventral area
BA basal amygdaloid nucleus
CA1 field CA1 of the hippocampus
CA2 field CA2 of the hippocampus
CA3 field CA3 of the hippocampus
CeA central amygdaloid nucleus
CL centrolateral thalamic nucleus
Cl claustrum
CM central medial thalamic nucleus
CPu caudate putamen (striatum)
D3V dorsal 3rd ventricle
DG dentate gyrus
DLG dorsal lateral geniculate nucleus
DMH dorsomedial hypothalamic nucleus
ec external capsule
En endopiriform nucleus
f fornix
fi fimbria of the hippocampus
GP globus pallidus
Hb habenula
ic internal capsule
In insular cortex
LA lateral amygdaloid nucleus
LP lateral posterior thalamic nucleus
LV lateral ventricle
MD mediodorsal thalamic nucleus
MePV med amy nucleus, post vent part
mfb medial forebrain bundle
mt mammillothalamic tract
opt optic tract
Pir piriform cortex
PPC posterior parietal cortex
Po posterior thalamic nuclear group
PrS presubiculum
PV paraventricular thalamic nucleus
Re reuniens thalamic nucleus
RS23 retrosplenial ctx area 23
RS29ab retrosplenial ctx area 29ab
RS29c retrosplenial ctx area 29c
RS30 retrosplenial ctx area 30
Rt reticular thalamic nucleus
S subiculum
SS somatosensory cortex
VMH ventromedial hypothalamic nucleus
VPL vent post lat thal nucleus
VPM vent post med thal nucleus
ZI zona incerta

## Plate 52 - Sinus: 5.71 mm

Interaural: 8.41 mm
Nissl section 110

Interaural: 8.37 mm
Parv section 110

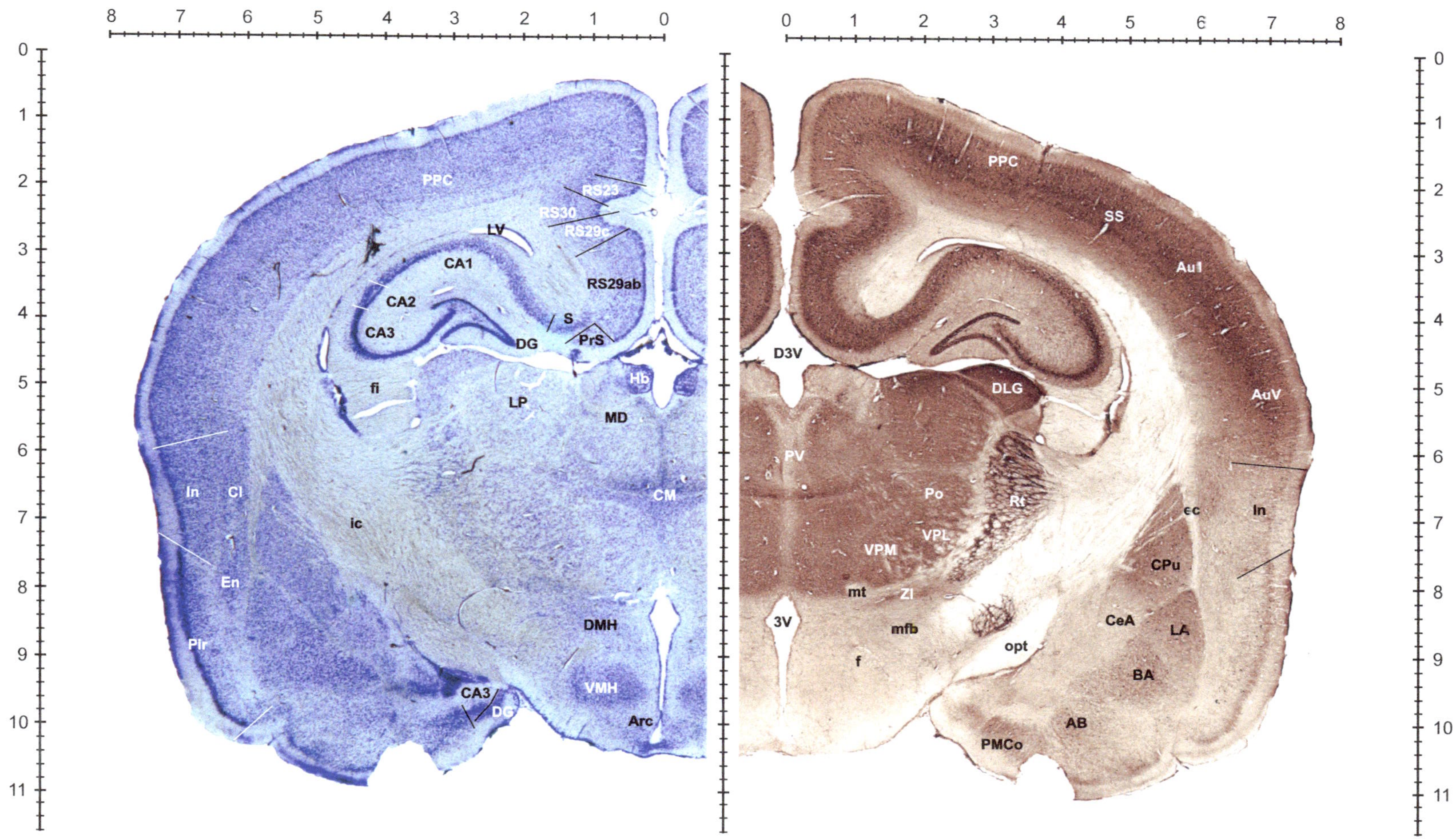


**3V** 3rd ventricle
**AB** accessory basal nucleus amygdala
**Arc** arcuate hypothalamic nucleus
**Au1** primary auditory cortex
**AuV** secondary auditory ctx ventral area
**BA** basal amygdaloid nucleus
**CA1** field CA1 of the hippocampus
**CA2** field CA2 of the hippocampus
**CA3** field CA3 of the hippocampus
**CeA** central amygdaloid nucleus
**Cl** claustrum
**CM** central medial thalamic nucleus
**CPu** caudate putamen (striatum)
**D3V** dorsal 3rd ventricle
**DG** dentate gyrus
**DLG** dorsal lateral geniculate nucleus
**DMH** dorsomedial hypothalamic nucleus
**ec** external capsule
**En** endopiriform nucleus
**f** fornix
**fi** fimbria of the hippocampus
**Hb** habenula
**ic** internal capsule
**In** insular cortex
**LA** lateral amygdaloid nucleus
**LP** lateral posterior thalamic nucleus
**LV** lateral ventricle
**MD** mediodorsal thalamic nucleus
**mfb** medial forebrain bundle
**mt** mammillothalamic tract
**opt** optic tract
**Pir** piriform cortex
**PMCo** postmed cortical amy
**Po** posterior thalamic nuclear group
**PPC** posterior parietal cortex
**PrS** presubiculum
**PV** paraventricular thalamic nucleus
**RS23** retrosplenial ctx area 23
**RS29ab** retrosplenial ctx area 29ab
**RS29c** retrosplenial ctx area 29c
**RS30** retrosplenial ctx area 30
**Rt** reticular thalamic nucleus
**S** subiculum
**SS** somatosensory cortex
**VMH** ventromedial hypothalamic nucleus
**VPL** vent post lat thal nucleus
**VPM** vent post med thal nucleus
**ZI** zona incerta

## Plate 53 - Sinus: 5.63 mm

Interaural: 8.33 mm
Nissl section 111

Interaural: 8.28 mm
AChE section 111

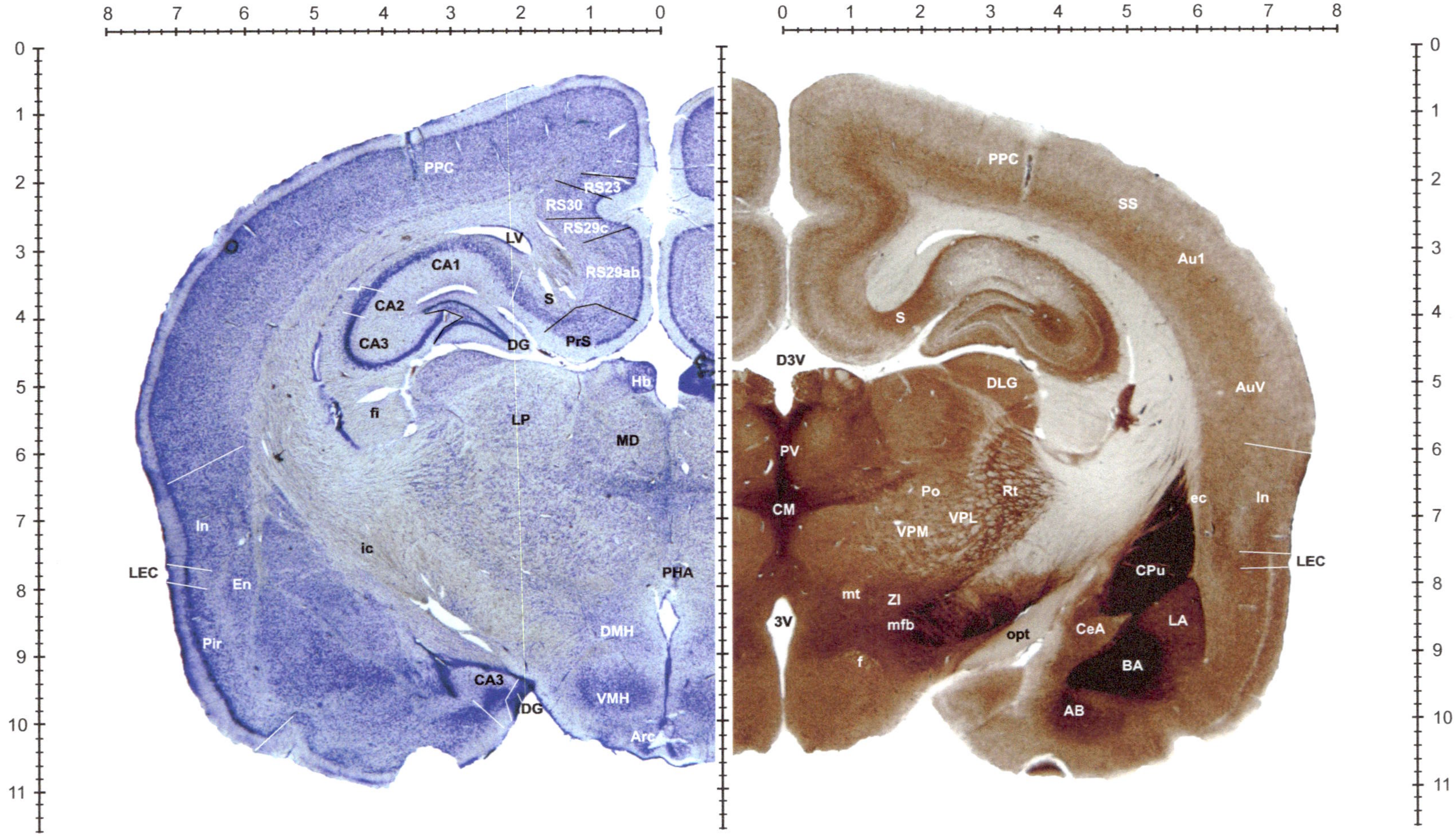


3V 3rd ventricle
AB accessory basal nucleus amygdala
Arc arcuate hypothalamic nucleus
Au1 primary auditory cortex
AuV secondary auditory ctx ventral area
BA basal amygdaloid nucleus
CA1 field CA1 of the hippocampus
CA2 field CA2 of the hippocampus
CA3 field CA3 of the hippocampus
CeA central amygdaloid nucleus
CM central medial thalamic nucleus
CPu caudate putamen (striatum)
D3V dorsal 3rd ventricle
DG dentate gyrus
DLG dorsal lateral geniculate nucleus
DMH dorsomedial hypothalamic nucleus
ec external capsule
En endopiriform nucleus
f fornix
fi fimbria of the hippocampus
Hb habenula
ic internal capsule
In insular cortex
LA lateral amygdaloid nucleus
LEC lateral entorhinal cortex
LP lateral posterior thalamic nucleus
LV lateral ventricle
MD mediodorsal thalamic nucleus
mfb medial forebrain bundle
mt mammillothalamic tract
opt optic tract
PHA posterior hypothalamic area
Pir piriform cortex
Po posterior thalamic nuclear group
PPC posterior parietal cortex
PrS presubiculum
PV paraventricular thalamic nucleus
RS23 retrosplenial ctx area 23
RS29ab retrosplenial ctx area 29ab
RS29c retrosplenial ctx area 29c
RS30 retrosplenial ctx area 30
Rt reticular thalamic nucleus
S subiculum
SS somatosensory cortex
VMH ventromedial hypothalamic nucleus
VPL vent post lat thal nucleus
VPM vent post med thal nucleus
ZI zona incerta

## Plate 54 - Sinus: 5.38 mm

Interaural: 8.08 mm
Nissl section 114

Interaural: 7.96 mm
AChE section 115

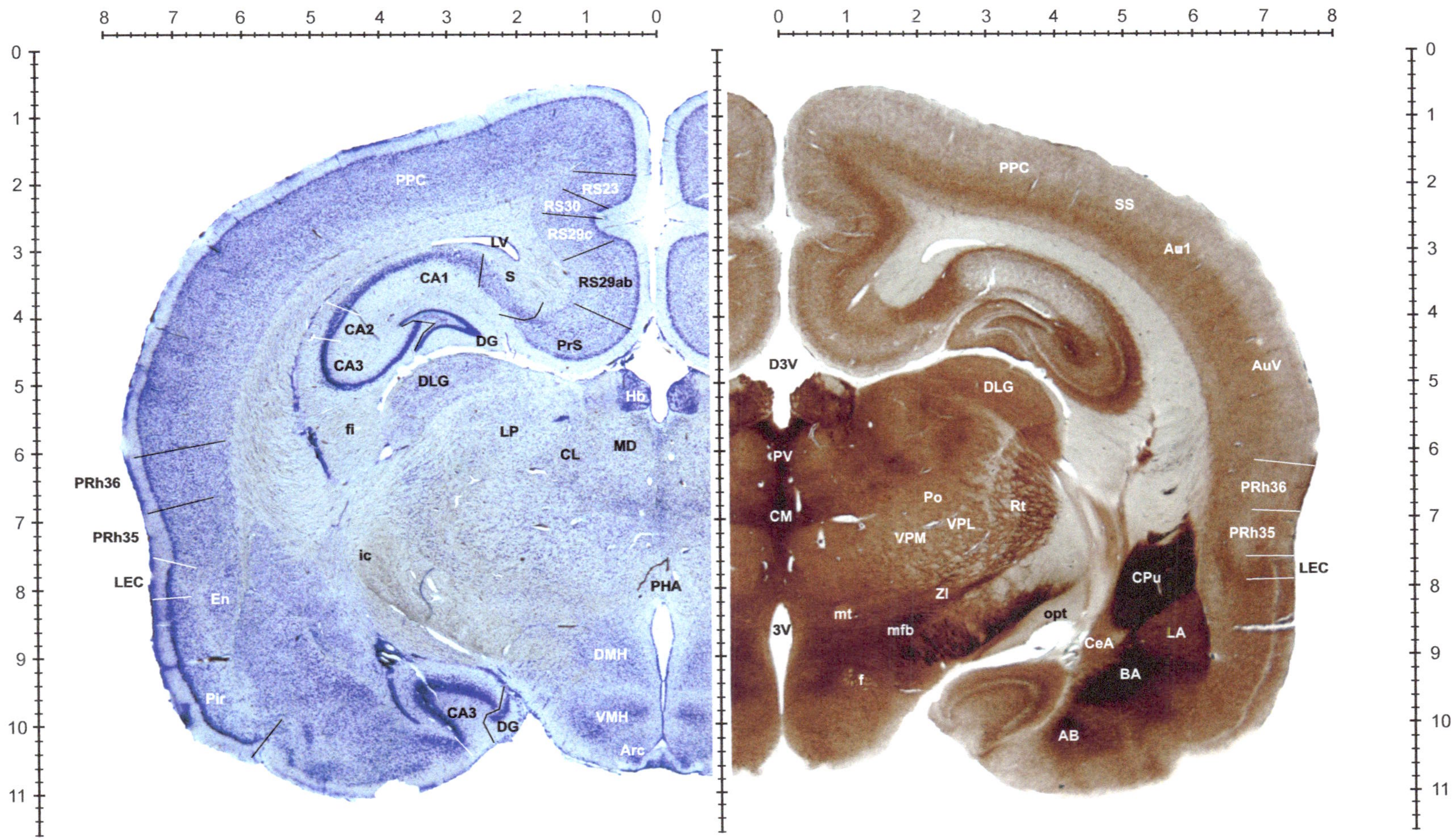


3V 3rd ventricle
AB accessory basal nucleus amygdala
Arc arcuate hypothalamic nucleus
Au1 primary auditory cortex
AuV secondary auditory ctx ventral area
BA basal amygdaloid nucleus
CA1 field CA1 of the hippocampus
CA2 field CA2 of the hippocampus
CA3 field CA3 of the hippocampus
CeA central amygdaloid nucleus
CL centrolateral thalamic nucleus
CM central medial thalamic nucleus
CPu caudate putamen (striatum)
D3V dorsal 3rd ventricle
DG dentate gyrus
DLG dorsal lateral geniculate nucleus
DMH dorsomedial hypothalamic nucleus
En endopiriform nucleus
f fornix
fi fimbria of the hippocampus
Hb habenula
ic internal capsule
LA lateral amygdaloid nucleus
LEC lateral entorhinal cortex
LP lateral posterior thalamic nucleus
LV lateral ventricle
MD mediodorsal thalamic nucleus
mfb medial forebrain bundle
mt mammillothalamic tract
opt optic tract
PHA posterior hypothalamic area
Pir piriform cortex
Po posterior thalamic nuclear group
PPC posterior parietal cortex
PRh35 perirhinal cortex, area 35
PRh36 perirhinal cortex, area 36
PrS presubiculum
PV paraventricular thalamic nucleus
RS23 retrosplenial ctx area 23
RS29ab retrosplenial ctx area 29ab
RS29c retrosplenial ctx area 29c
RS30 retrosplenial ctx area 30
Rt reticular thalamic nucleus
S subiculum
SS somatosensory cortex
VMH ventromedial hypothalamic nucleus
VPL vent post lat thal nucleus
VPM vent post med thal nucleus
ZI zona incerta

## Plate 55 - Sinus: 5.13 mm

Interaural: 7.83 mm
Nissl section 117

Interaural: 7.87 mm
Parv section 116

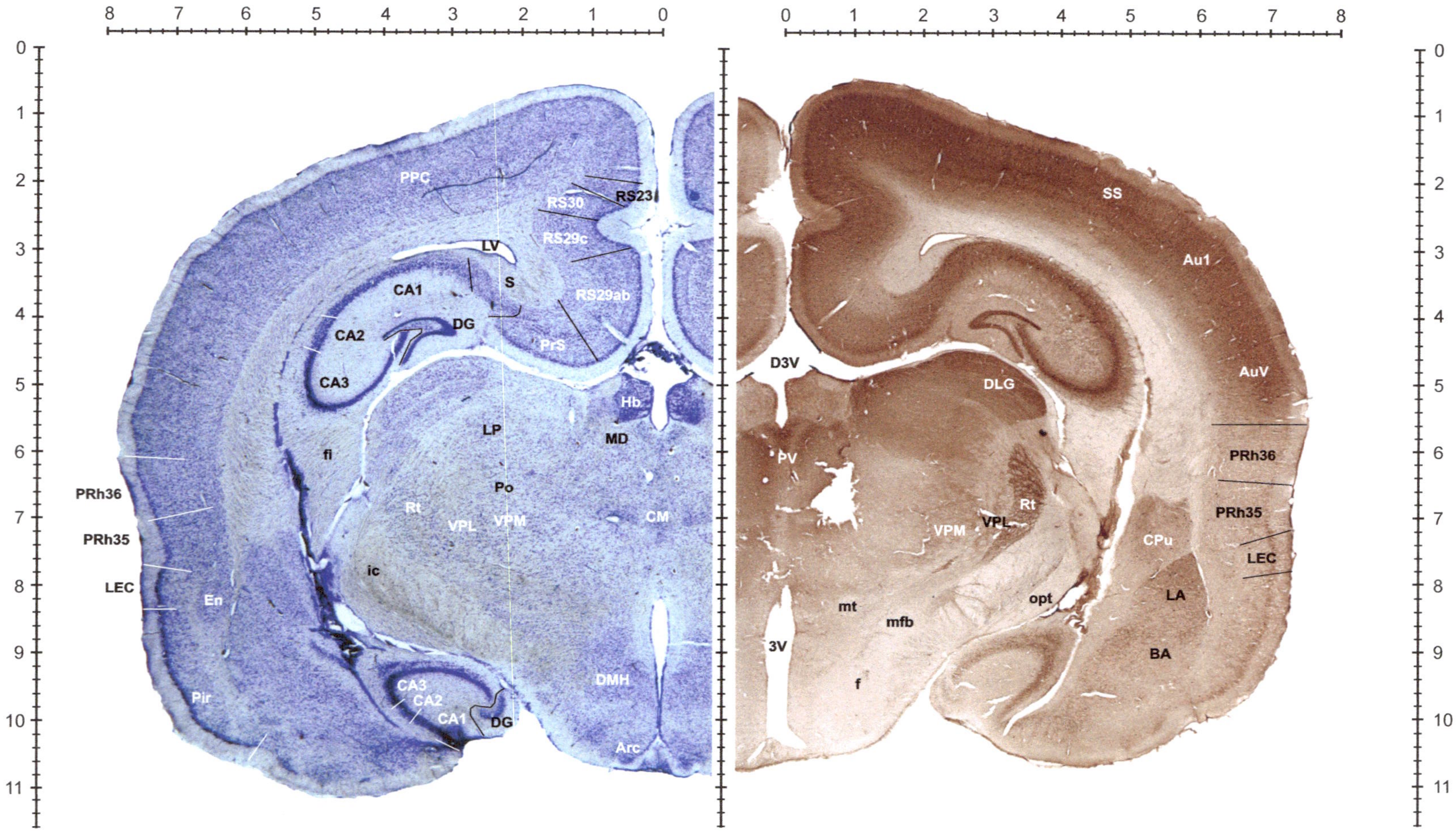


**3V** 3rd ventricle
**Arc** arcuate hypothalamic nucleus
**Au1** primary auditory cortex
**AuV** secondary auditory ctx ventral area
**BA** basal amygdaloid nucleus
**CA1** field CA1 of the hippocampus
**CA2** field CA2 of the hippocampus
**CA3** field CA3 of the hippocampus
**CM** central medial thalamic nucleus
**CPu** caudate putamen (striatum)
**D3V** dorsal 3rd ventricle
**DG** dentate gyrus
**DLG** dorsal lateral geniculate nucleus
**DMH** dorsomedial hypothalamic nucleus
**En** endopiriform nucleus
**f** fornix
**fi** fimbria of the hippocampus
**Hb** habenula
**ic** internal capsule
**LA** lateral amygdaloid nucleus
**LEC** lateral entorhinal cortex
**LP** lateral posterior thalamic nucleus
**LV** lateral ventricle
**MD** mediodorsal thalamic nucleus
**mfb** medial forebrain bundle
**mt** mammillothalamic tract
**opt** optic tract
**Pir** piriform cortex
**Po** posterior thalamic nuclear group
**PPC** posterior parietal cortex
**PRh35** perirhinal cortex, area 35
**PRh36** perirhinal cortex, area 36
**PrS** presubiculum
**PV** paraventricular thalamic nucleus
**RS23** retrosplenial ctx area 23
**RS29ab** retrosplenial ctx area 29ab
**RS29c** retrosplenial ctx area 29c
**RS30** retrosplenial ctx area 30
**Rt** reticular thalamic nucleus
**S** subiculum
**SS** somatosensory cortex
**VPL** vent post lat thal nucleus
**VPM** vent post med thal nucleus

## Plate 56 - Sinus: 5.05 mm

Interaural: 7.75 mm
Nissl section 118

Interaural: 7.79 mm
AChE section 117

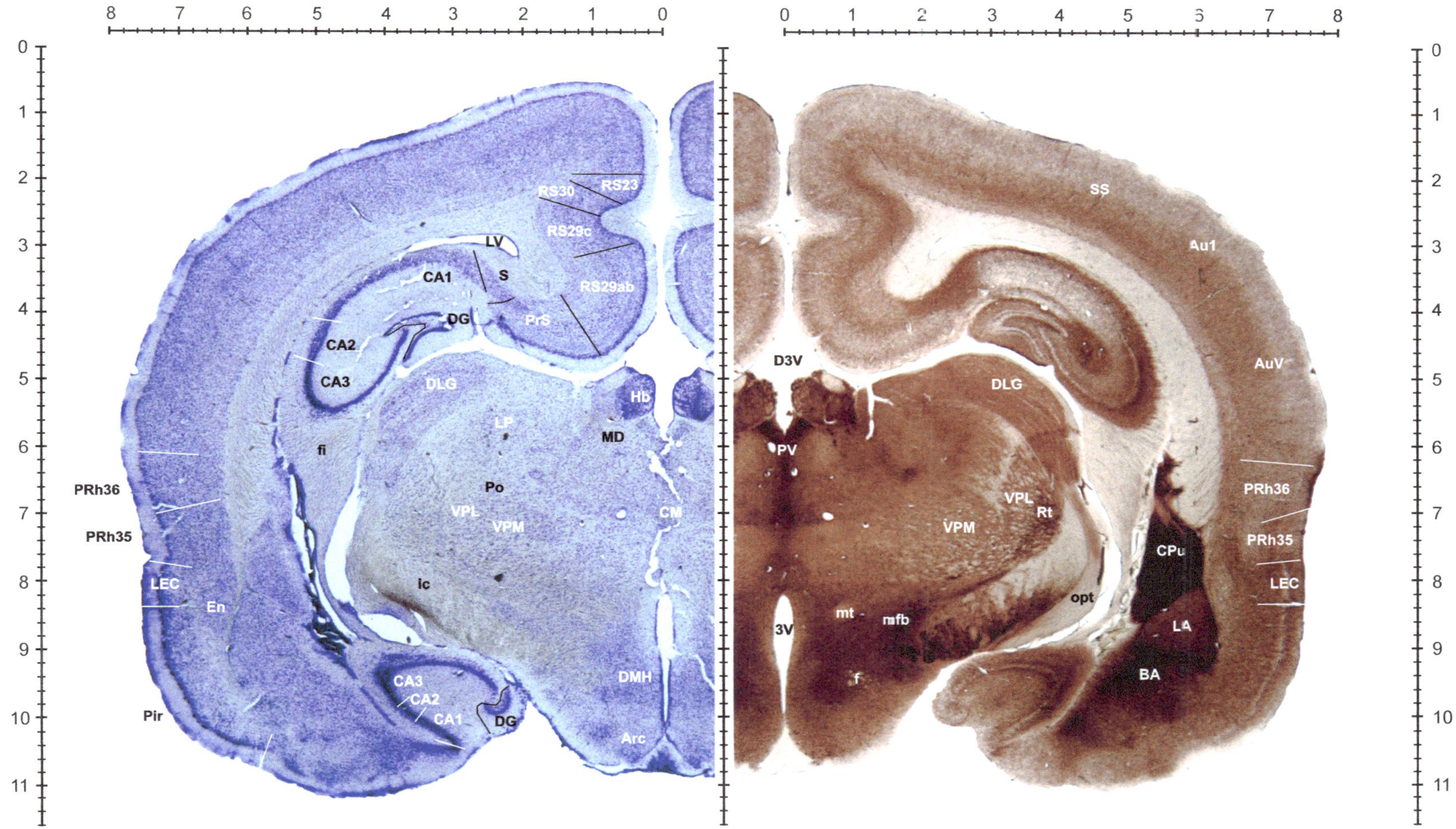


3V 3rd ventricle
Arc arcuate hypothalamic nucleus
Au1 primary auditory cortex
AuV secondary auditory ctx ventral area
BA basal amygdaloid nucleus
CA1 field CA1 of the hippocampus
CA2 field CA2 of the hippocampus
CA3 field CA3 of the hippocampus
CM central medial thalamic nucleus
CPu caudate putamen (striatum)
D3V dorsal 3rd ventricle
DG dentate gyrus
DLG dorsal lateral geniculate nucleus
DMH dorsomedial hypothalamic nucleus
En endopiriform nucleus
f fornix
fi fimbria of the hippocampus
Hb habenula
ic internal capsule
LA lateral amygdaloid nucleus
LEC lateral entorhinal cortex
LP lateral posterior thalamic nucleus
LV lateral ventricle
MD mediodorsal thalamic nucleus
mfb medial forebrain bundle
mt mammillothalamic tract
opt optic tract
Pir piriform cortex
Po posterior thalamic nuclear group
PRh35 perirhinal cortex, area 35
PRh36 perirhinal cortex, area 36
PrS presubiculum
PV paraventricular thalamic nucleus
RS23 retrosplenial ctx area 23
RS29ab retrosplenial ctx area 29ab
RS29c retrosplenial ctx area 29c
RS30 retrosplenial ctx area 30
Rt reticular thalamic nucleus
S subiculum
SS somatosensory cortex
VPL vent post lat thal nucleus
VPM vent post med thal nucleus

**Plate 57 - Sinus: 4.89 mm**

Interaural: 7.59 mm
Nissl section 120

Interaural: 7.46 mm
AChE section 121

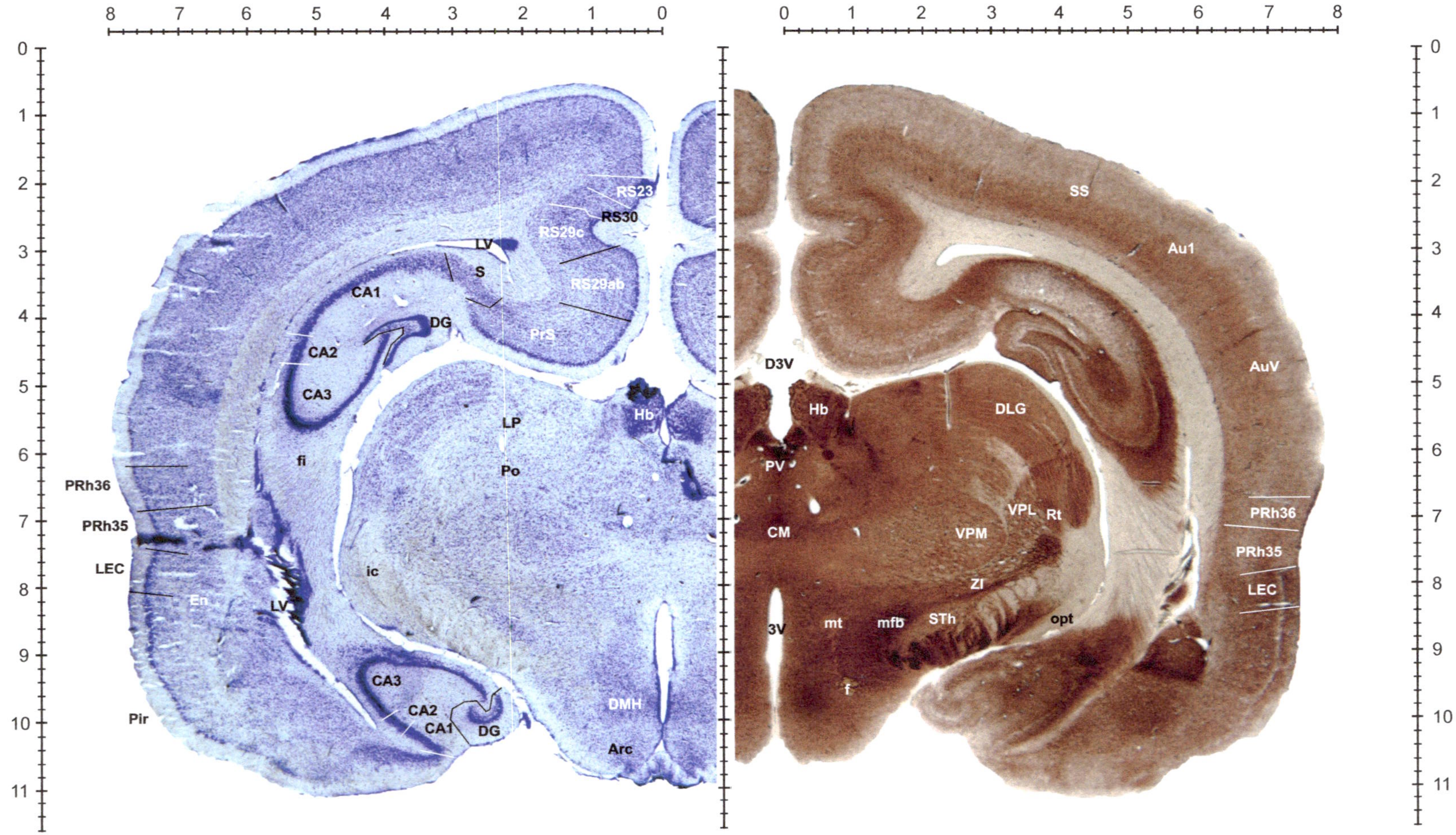


**3V** 3rd ventricle
**Arc** arcuate hypothalamic nucleus
**Au1** primary auditory cortex
**AuV** secondary auditory ctx ventral area
**CA1** field CA1 of the hippocampus
**CA2** field CA2 of the hippocampus
**CA3** field CA3 of the hippocampus
**CM** central medial thalamic nucleus
**D3V** dorsal 3rd ventricle
**DG** dentate gyrus
**DLG** dorsal lateral geniculate nucleus
**DMH** dorsomedial hypothalamic nucleus
**En** endopiriform nucleus
**f** fornix
**fi** fimbria of the hippocampus
**Hb** habenula
**ic** internal capsule
**LEC** lateral entorhinal cortex
**LP** lateral posterior thalamic nucleus
**LV** lateral ventricle
**mfb** medial forebrain bundle
**mt** mammillothalamic tract
**opt** optic tract
**Pir** piriform cortex
**Po** posterior thalamic nuclear group
**PRh35** perirhinal cortex, area 35
**PRh36** perirhinal cortex, area 36
**PrS** presubiculum
**PV** paraventricular thalamic nucleus
**RS23** retrosplenial ctx area 23
**RS29ab** retrosplenial ctx area 29ab
**RS29c** retrosplenial ctx area 29c
**RS30** retrosplenial ctx area 30
**Rt** reticular thalamic nucleus
**S** subiculum
**SS** somatosensory cortex
**STh** subthalamic nucleus
**VPL** vent post lat thal nucleus
**VPM** vent post med thal nucleus
**ZI** zona incerta

## Plate 58 - Sinus: 4.72 mm

Interaural: 7.42 mm
Nissl section 122

Interaural: 7.38 mm
Parv section 122

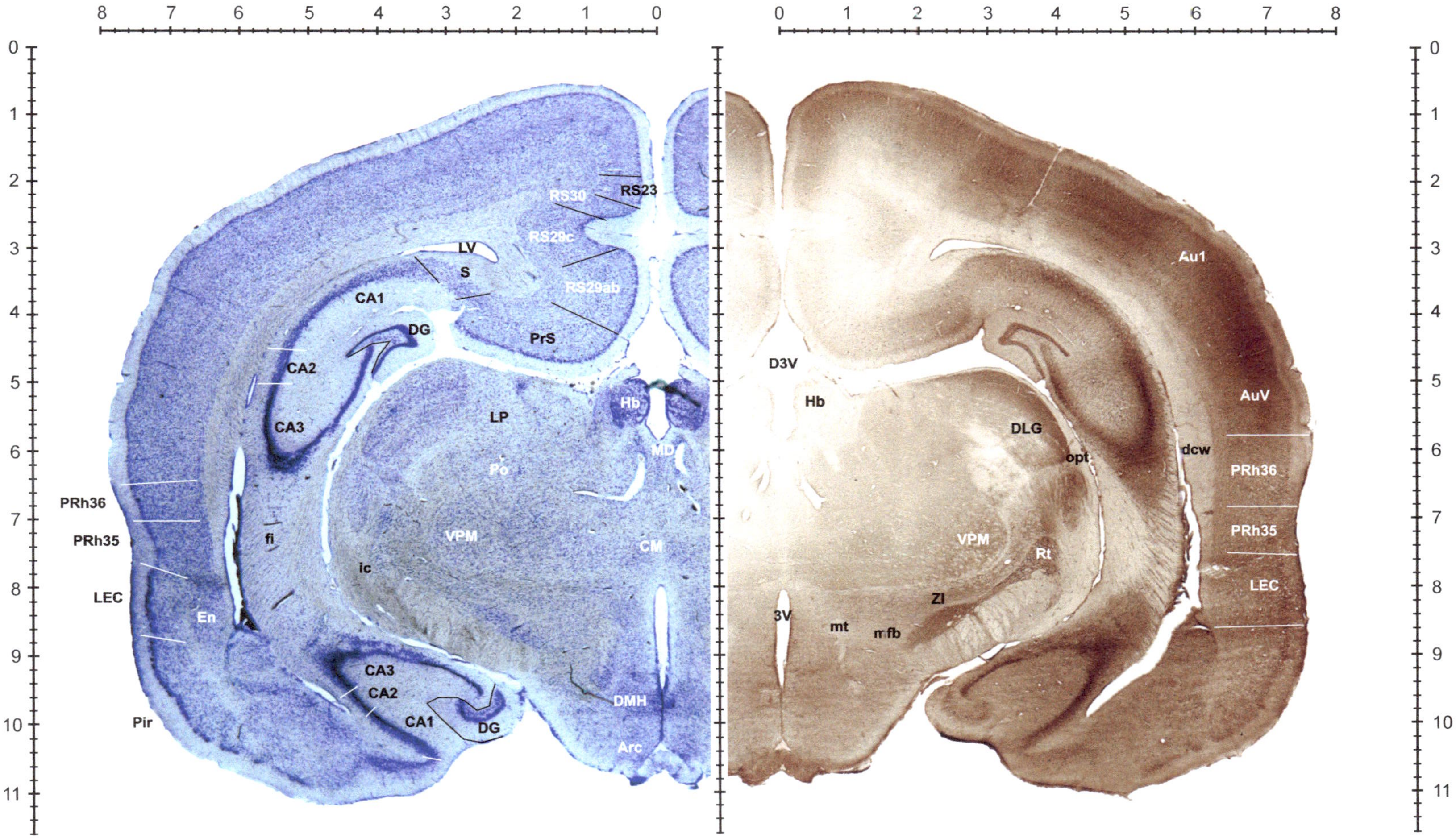


**3V** 3rd ventricle
**Arc** arcuate hypothalamic nucleus
**Au1** primary auditory cortex
**AuV** secondary auditory ctx ventral area
**CA1** field CA1 of the hippocampus
**CA2** field CA2 of the hippocampus
**CA3** field CA3 of the hippocampus
**CM** central medial thalamic nucleus
**D3V** dorsal 3rd ventricle
**dcw** deep cerebral white matter
**DG** dentate gyrus
**DLG** dorsal lateral geniculate nucleus
**DMH** dorsomedial hypothalamic nucleus
**En** endopiriform nucleus
**fi** fimbria of the hippocampus
**Hb** habenula
**ic** internal capsule
**LEC** lateral entorhinal cortex
**LP** lateral posterior thalamic nucleus
**LV** lateral ventricle
**MD** mediodorsal thalamic nucleus
**mfb** medial forebrain bundle
**mt** mammillothalamic tract
**opt** optic tract
**Pir** piriform cortex
**Po** posterior thalamic nuclear group
**PRh35** perirhinal cortex, area 35
**PRh36** perirhinal cortex, area 36
**PrS** presubiculum
**RS23** retrosplenial ctx area 23
**RS29ab** retrosplenial ctx area 29ab
**RS29c** retrosplenial ctx area 29c
**RS30** retrosplenial ctx area 30
**Rt** reticular thalamic nucleus
**S** subiculum
**VPM** vent post med thal nucleus
**ZI** zona incerta

**Plate 59 - Sinus: 4.56 mm**

Interaural: 7.26 mm
Nissl section 124

Interaural: 7.22 mm
AChE section 124

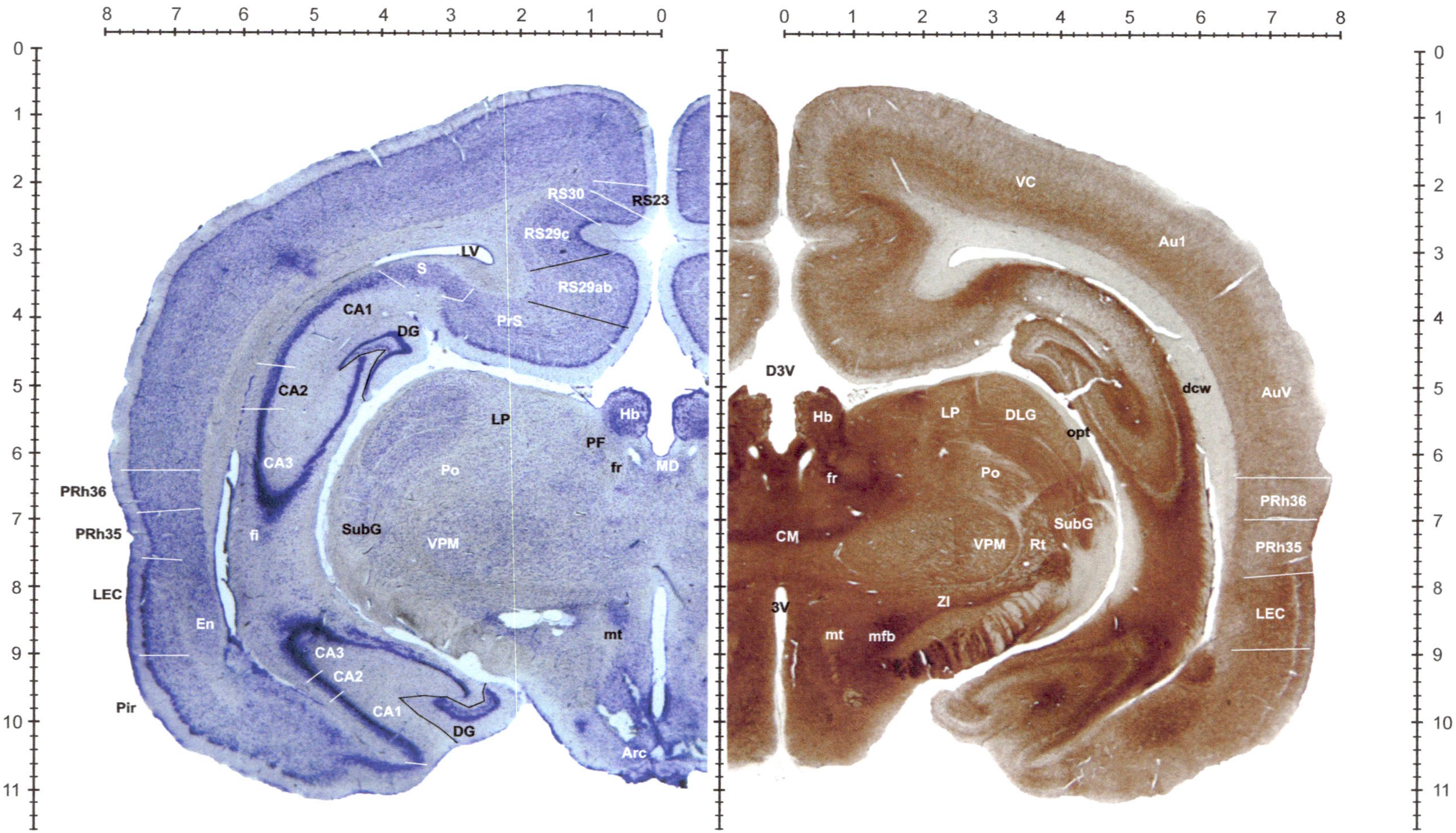


**3V** 3rd ventricle
**Arc** arcuate hypothalamic nucleus
**Au1** primary auditory cortex
**AuV** secondary auditory ctx ventral area
**CA1** field CA1 of the hippocampus
**CA2** field CA2 of the hippocampus
**CA3** field CA3 of the hippocampus
**CM** central medial thalamic nucleus
**D3V** dorsal 3rd ventricle
**dcw** deep cerebral white matter
**DG** dentate gyrus
**DLG** dorsal lateral geniculate nucleus
**En** endopiriform nucleus
**fi** fimbria of the hippocampus
**fr** fasciculus retroflexus
**Hb** habenula
**LEC** lateral entorhinal cortex
**LP** lateral posterior thalamic nucleus
**LV** lateral ventricle
**MD** mediodorsal thalamic nucleus
**mfb** medial forebrain bundle
**mt** mammillothalamic tract
**opt** optic tract
**PF** parafascicular thalamic nucleus
**Pir** piriform cortex
**Po** posterior thalamic nuclear group
**PRh35** perirhinal cortex, area 35
**PRh36** perirhinal cortex, area 36
**PrS** presubiculum
**RS23** retrosplenial ctx area 23
**RS29ab** retrosplenial ctx area 29ab
**RS29c** retrosplenial ctx area 29c
**RS30** retrosplenial ctx area 30
**Rt** reticular thalamic nucleus
**S** subiculum
**SubG** subgeniculate nu of thal
**VC** visual cortex
**VPM** vent post med thal nucleus
**ZI** zona incerta

## Plate 60 - Sinus: 4.48 mm

Interaural: 7.18 mm

Nissl section 125

Interaural: 7.14 mm

Parv section 125

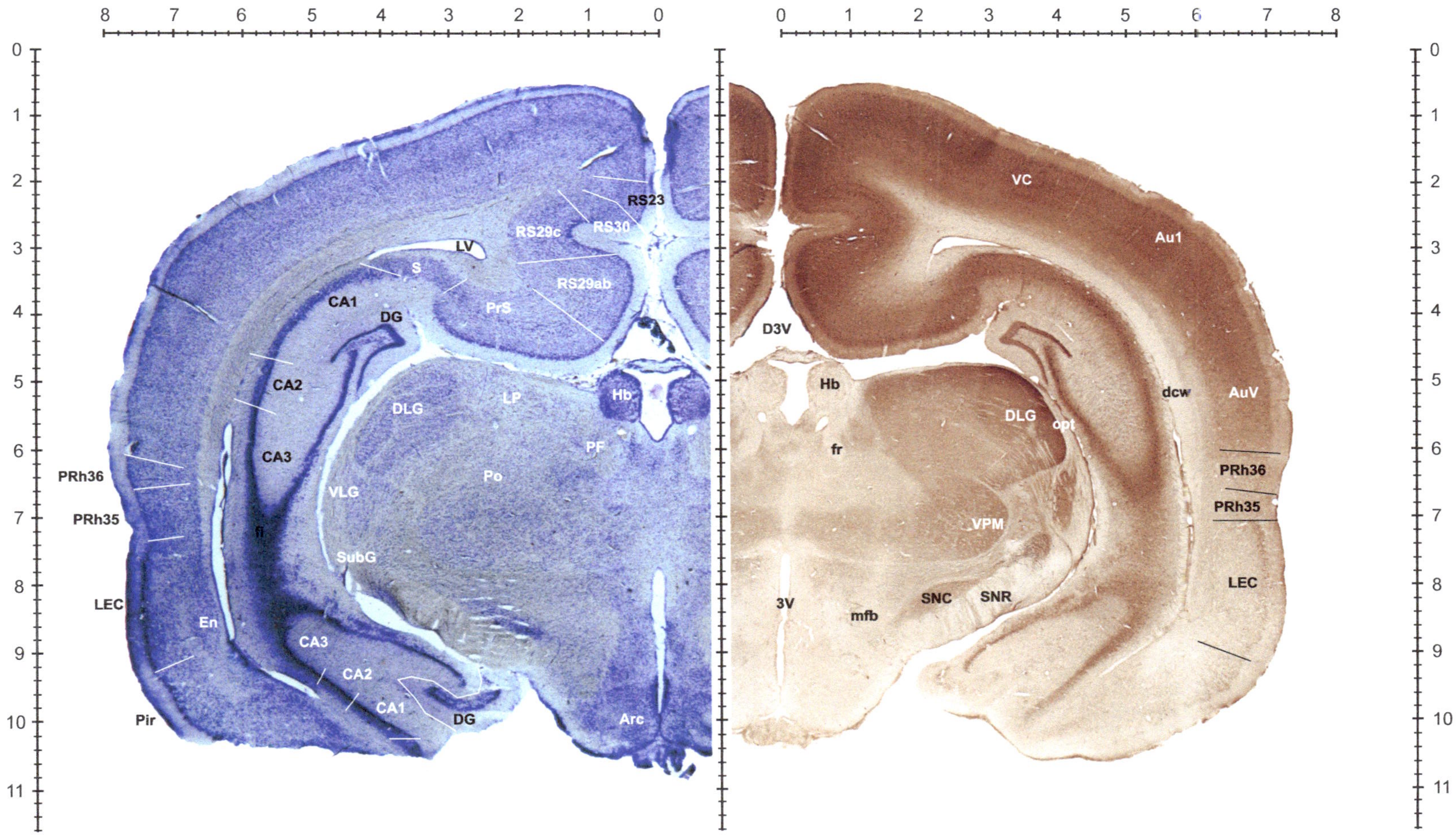


**3V** 3rd ventricle
**Arc** arcuate hypothalamic nucleus
**Au1** primary auditory cortex
**AuV** secondary auditory ctx ventral area
**CA1** field CA1 of the hippocampus
**CA2** field CA2 of the hippocampus
**CA3** field CA3 of the hippocampus
**D3V** dorsal 3rd ventricle
**dcw** deep cerebral white matter
**DG** dentate gyrus
**DLG** dorsal lateral geniculate nucleus
**En** endopiriform nucleus
**fr** fasciculus retroflexus
**Hb** habenula
**LEC** lateral entorhinal cortex
**LP** lateral posterior thalamic nucleus
**LV** lateral ventricle
**mfb** medial forebrain bundle
**opt** optic tract
**PF** parafascicular thalamic nucleus
**Pir** piriform cortex
**Po** posterior thalamic nuclear group
**PRh35** perirhinal cortex, area 35
**PRh36** perirhinal cortex, area 36
**PrS** presubiculum
**RS23** retrosplenial ctx area 23
**RS29ab** retrosplenial ctx area 29ab
**RS29c** retrosplenial ctx area 29c
**RS30** retrosplenial ctx area 30
**S** subiculum
**SNC** substantia nigra, compact part
**SNR** substantia nigra, reticular part
**SubG** subgeniculate nu of thal
**VC** visual cortex
**VLG** ventrolateral geniculate nucleus
**VPM** vent post med thal nucleus

## Plate 61 - Sinus: 4.31 mm

Interaural: 7.01 mm
Nissl section 127

Interaural: 6.97 mm
AChE section 127

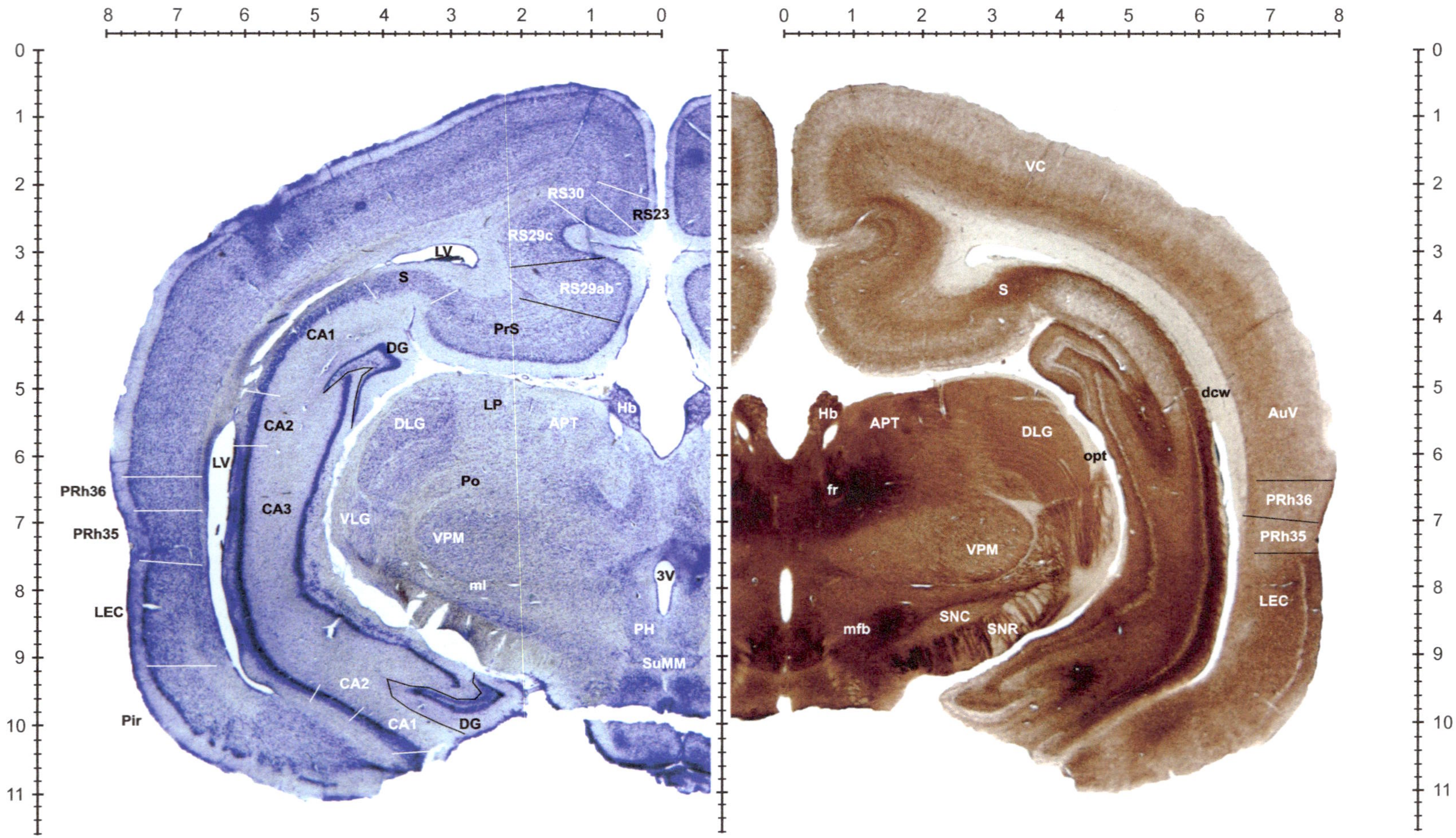


**3V** 3rd ventricle
**APT** anterior pretectal nucleus
**AuV** secondary auditory ctx ventral area
**CA1** field CA1 of the hippocampus
**CA2** field CA2 of the hippocampus
**CA3** field CA3 of the hippocampus
**dcw** deep cerebral white matter
**DG** dentate gyrus
**DLG** dorsal lateral geniculate nucleus
**fr** fasciculus retroflexus
**Hb** habenula
**LEC** lateral entorhinal cortex
**LP** lateral posterior thalamic nucleus
**LV** lateral ventricle
**ml** medial lemniscus
**mfb** medial forebrain bundle
**opt** optic tract
**PH** posterior hypothalamic nucleus
**Pir** piriform cortex
**Po** posterior thalamic nuclear group
**PRh35** perirhinal cortex, area 35
**PRh36** perirhinal cortex, area 36
**PrS** presubiculum
**RS23** retrosplenial ctx area 23
**RS29ab** retrosplenial ctx area 29ab
**RS29c** retrosplenial ctx area 29c
**RS30** retrosplenial ctx area 30
**S** subiculum
**SNC** substantia nigra, compact part
**SNR** substantia nigra, reticular part
**SuMM** supramammillary nu, medial part
**VC** visual cortex
**VLG** ventrolateral geniculate nucleus
**VPM** vent post med thal nucleus

## Plate 62 - Sinus: 4.23 mm

Interaural: 6.93 mm
Nissl section 128

Interaural: 6.81 mm
AChE section 129

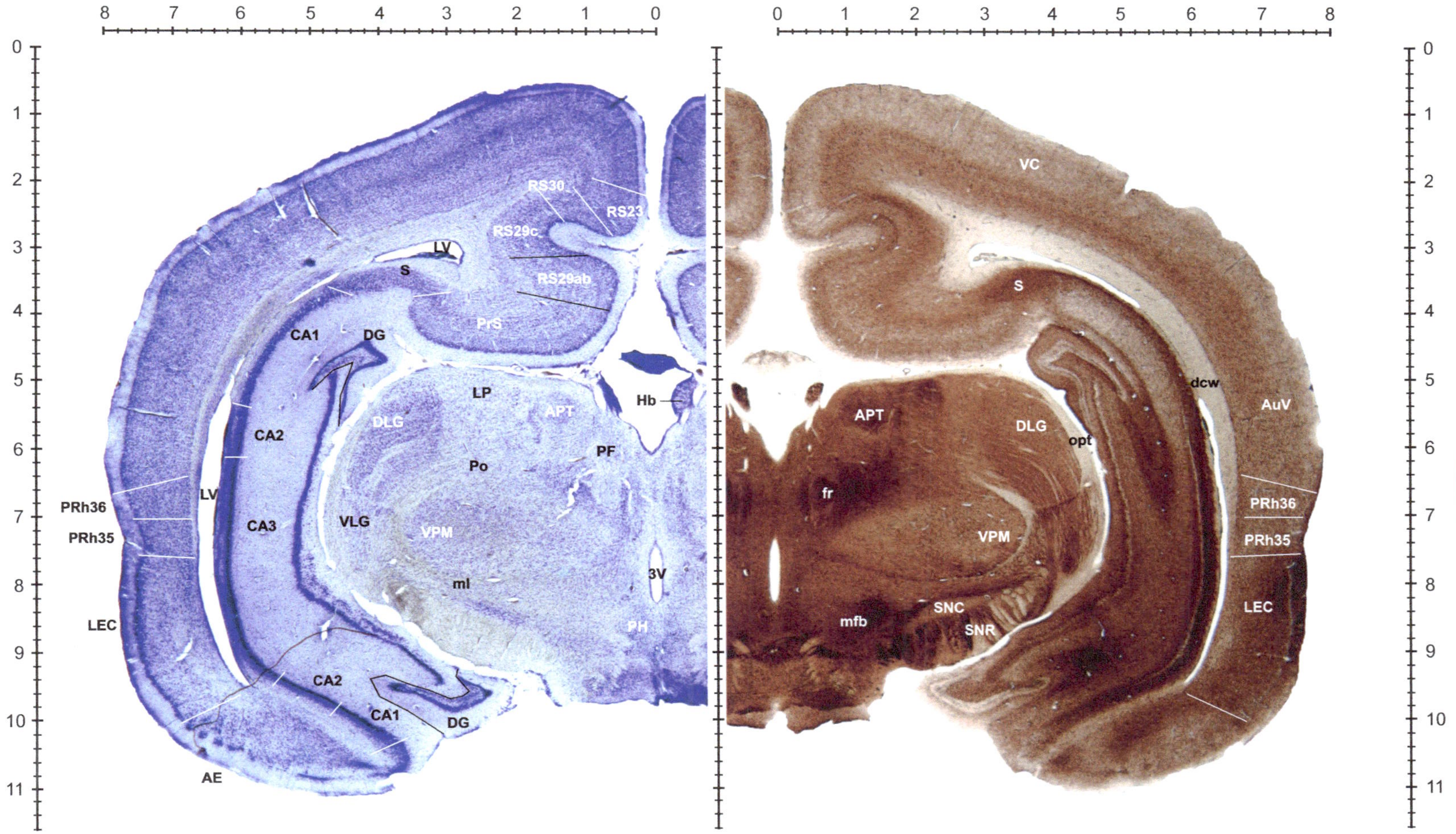


3V 3rd ventricle
AE amygdalo-entorhinal transition area
APT anterior pretectal nucleus
AuV secondary auditory ctx ventral area
CA1 field CA1 of the hippocampus
CA2 field CA2 of the hippocampus
CA3 field CA3 of the hippocampus
dcw deep cerebral white matter
DG dentate gyrus
DLG dorsal lateral geniculate nucleus
fr fasciculus retroflexus
Hb habenula
LEC lateral entorhinal cortex
LP lateral posterior thalamic nucleus
LV lateral ventricle
mfb medial forebrain bundle
ml medial lemniscus
opt optic tract
PH posterior hypothalamic nucleus
PF parafascicular thalamic nucleus
Po posterior thalamic nuclear group
PRh35 perirhinal cortex, area 35
PRh36 perirhinal cortex, area 36
PrS presubiculum
RS23 retrosplenial ctx area 23
RS29ab retrosplenial ctx area 29ab
RS29c retrosplenial ctx area 29c
RS30 retrosplenial ctx area 30
S subiculum
SNC substantia nigra, compact part
SNR substantia nigra, reticular part
VC visual cortex
VLG ventrolateral geniculate nucleus
VPM vent post med thal nucleus

**Plate 63 - Sinus: 4.15 mm**

Interaural: 6.85 mm
Nissl section 129

Interaural: 6.73 mm
AChE section 130

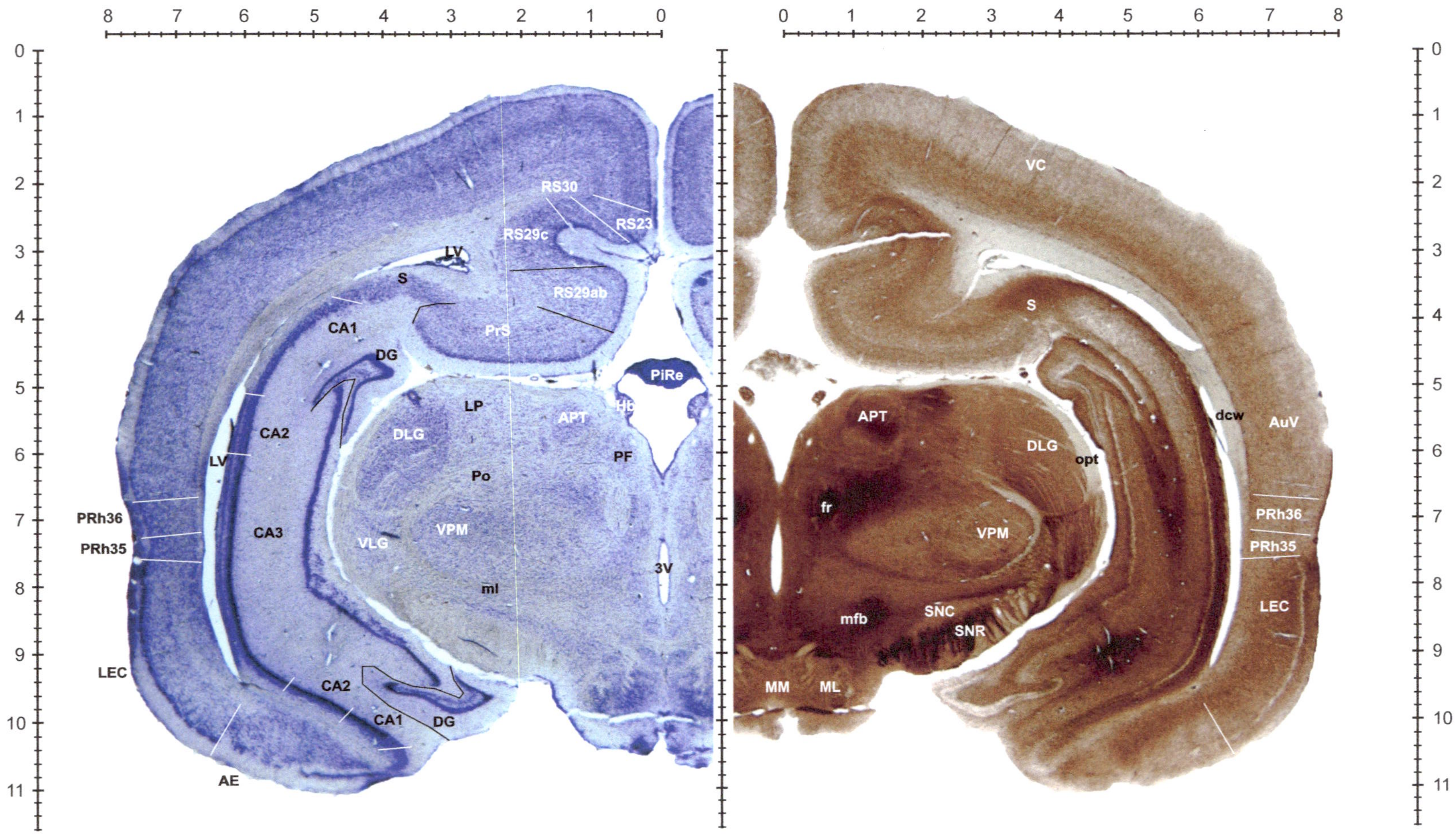


**3V** 3rd ventricle
**AE** amygdalo-entorhinal transition area
**APT** anterior pretectal nucleus
**AuV** secondary auditory ctx ventral area
**CA1** field CA1 of the hippocampus
**CA2** field CA2 of the hippocampus
**CA3** field CA3 of the hippocampus
**dcw** deep cerebral white matter
**DG** dentate gyrus
**DLG** dorsal lateral geniculate nucleus
**fr** fasciculus retroflexus
**Hb** habenula
**LEC** lateral entorhinal cortex
**LP** lateral posterior thalamic nucleus
**LV** lateral ventricle
**mfb** medial forebrain bundle
**ML** med mamm nu, lat part
**ml** medial lemniscus
**MM** med mamm nu, med part
**opt** optic tract
**PF** parafascicular thalamic nucleus
**PiRe** pineal recess
**Po** posterior thalamic nuclear group
**PRh35** perirhinal cortex, area 35
**PRh36** perirhinal cortex, area 36
**PrS** presubiculum
**RS23** retrosplenial ctx area 23
**RS29ab** retrosplenial ctx area 29ab
**RS29c** retrosplenial ctx area 29c
**RS30** retrosplenial ctx area 30
**S** subiculum
**SNC** substantia nigra, compact part
**SNR** substantia nigra, reticular part
**VC** visual cortex
**VLG** ventrolateral geniculate nucleus
**VPM** vent post med thal nucleus

**Plate 64 - Sinus: 3.90 mm**

Interaural: 6.60 mm
Nissl section 132

Interaural: 6.56 mm
AChE section 132

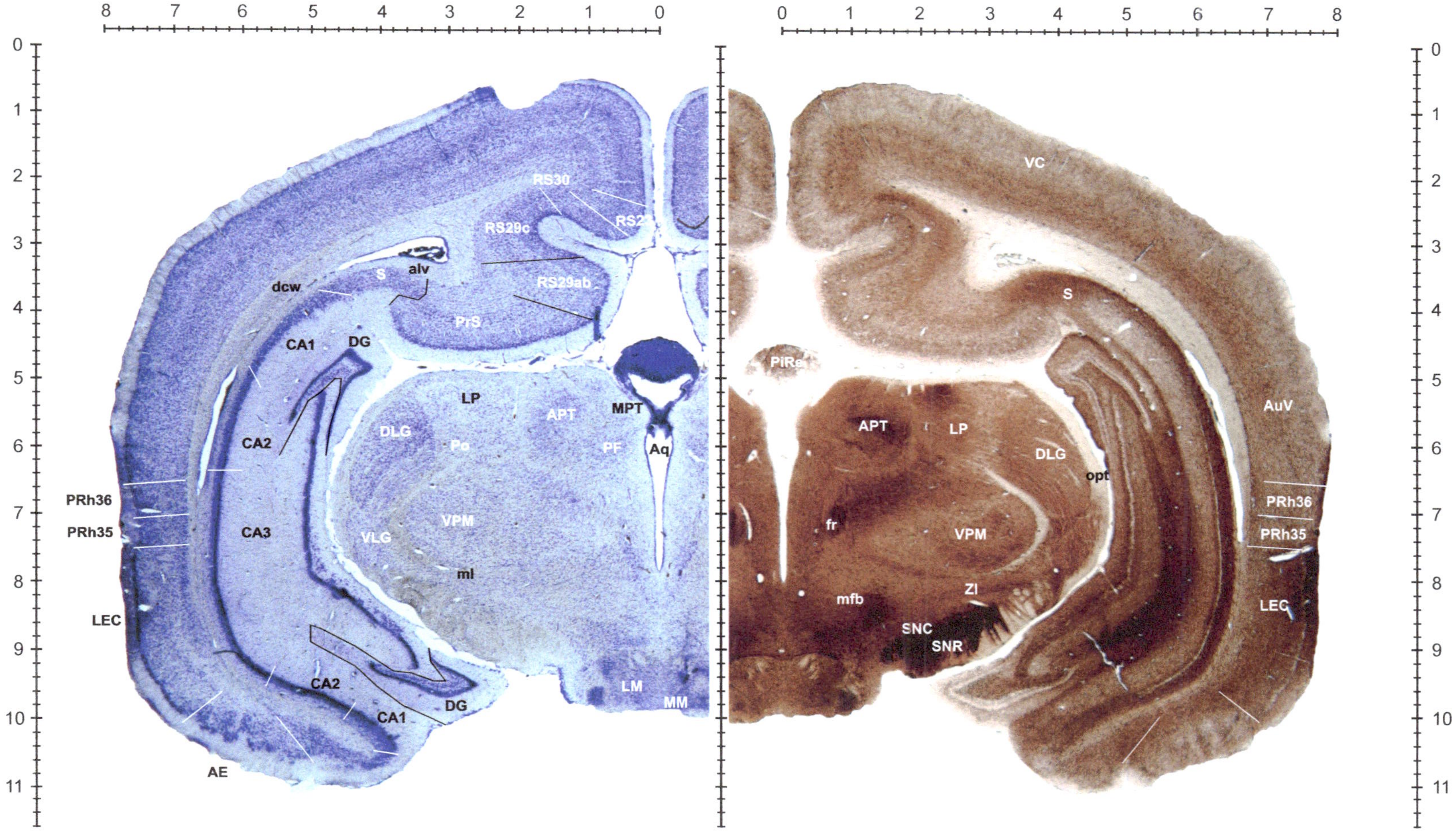


**AE** amygdalo-entorhinal transition area
**alv** alveus of the hippocampus
**APT** anterior pretectal nucleus
**Aq** aqueduct
**AuV** secondary auditory ctx ventral area
**CA1** field CA1 of the hippocampus
**CA2** field CA2 of the hippocampus
**CA3** field CA3 of the hippocampus
**dcw** deep cerebral white matter
**DG** dentate gyrus
**DLG** dorsal lateral geniculate nucleus
**fr** fasciculus retroflexus
**LEC** lateral entorhinal cortex
**LM** lateral mammillary nucleus
**LP** lateral posterior thalamic nucleus
**mfb** medial forebrain bundle
**ml** medial lemniscus
**MM** med mamm nucleus, med part
**MPT** medial pretectal nucleus
**opt** optic tract
**PF** parafascicular thalamic nucleus
**PiRe** pineal recess
**Po** posterior thalamic nuclear group
**PRh35** perirhinal cortex, area 35
**PRh36** perirhinal cortex, area 36
**PrS** presubiculum
**RS23** retrosplenial ctx area 23
**RS29ab** retrosplenial ctx area 29ab
**RS29c** retrosplenial ctx area 29c
**RS30** retrosplenial ctx area 30
**S** subiculum
**SNC** substantia nigra, compact part
**SNR** substantia nigra, reticular part
**VC** visual cortex
**VLG** ventrolateral geniculate nucleus
**VPM** vent post med thal nucleus
**ZI** zona incerta

## Plate 65 - Sinus: 3.74 mm

Interaural: 6.44 mm
Nissl section 134

Interaural: 6.40 mm
Parv section 134

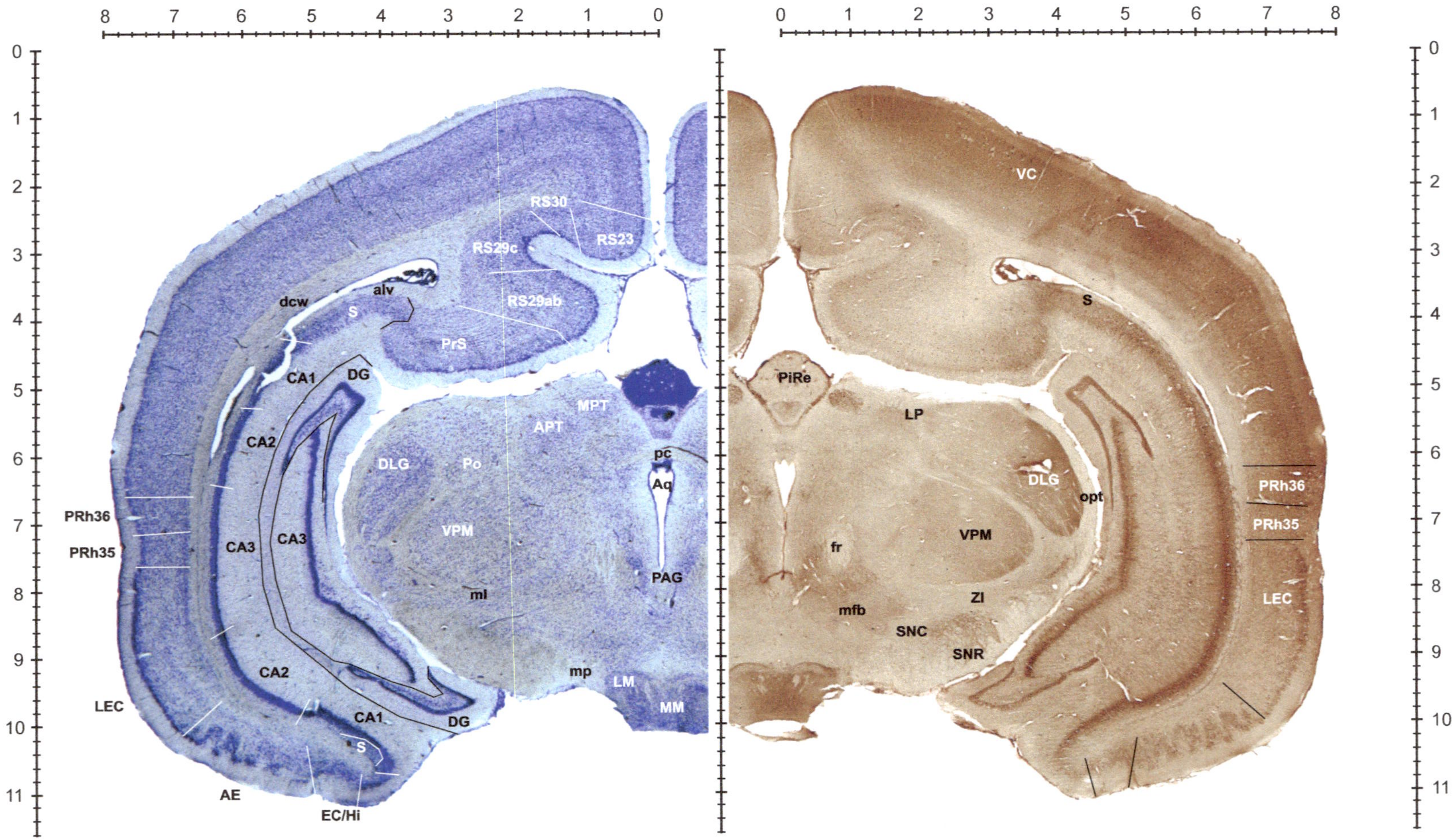


**AE** amygdalo-entorhinal transition area
**alv** alveus of the hippocampus
**APT** anterior pretectal nucleus
**Aq** aqueduct
**CA1** field CA1 of the hippocampus
**CA2** field CA2 of the hippocampus
**CA3** field CA3 of the hippocampus
**dcw** deep cerebral white matter
**DG** dentate gyrus
**DLG** dorsal lateral geniculate nucleus
**EC/Hi** entorhinal cx-hippocampal transition
**fr** fasciculus retroflexus
**LEC** lateral entorhinal cortex
**LM** lateral mammillary nucleus
**LP** lat post thal nucleus
**mfb** medial forebrain bundle
**ml** medial lemniscus
**MM** med mamm nucleus, med part
**mp** mammillary peduncle
**MPT** medial pretectal nucleus
**opt** optic tract
**PAG** periaqueductal gray
**pc** posterior commissure
**PiRe** pineal recess
**Po** posterior thalamic nuclear group
**PRh35** perirhinal cortex, area 35
**PRh36** perirhinal cortex, area 36
**PrS** presubiculum
**RS23** retrosplenial ctx area 23
**RS29ab** retrosplenial ctx area 29ab
**RS29c** retrosplenial ctx area 29c
**RS30** retrosplenial ctx area 30
**S** subiculum
**SNC** substantia nigra, compact part
**SNR** substantia nigra, reticular part
**VC** visual cortex
**VPM** vent post med thal nucleus
**ZI** zona incerta

## Plate 66 - Sinus: 3.58 mm

Interaural: 6.28 mm
Nissl section 136

Interaural: 6.32 mm
AChE section 135

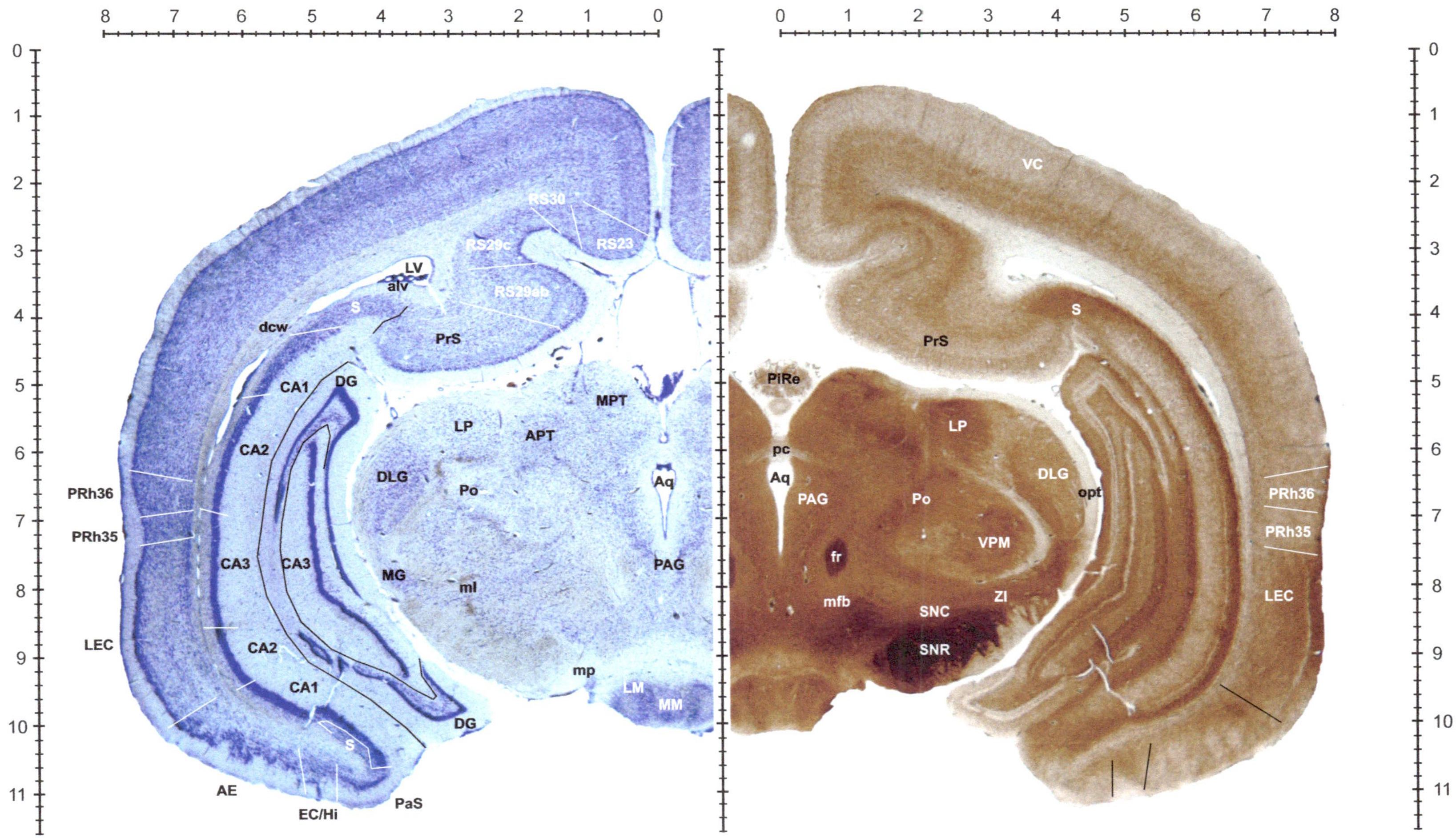


**AE** amygdalo-entorhinal transition area
**alv** alveus of the hippocampus
**APT** anterior pretectal nucleus
**Aq** aqueduct
**CA1** field CA1 of the hippocampus
**CA2** field CA2 of the hippocampus
**CA3** field CA3 of the hippocampus
**dcw** deep cerebral white matter
**DG** dentate gyrus
**DLG** dorsal lateral geniculate nucleus
**EC/Hi** entorhinal cx-hippocampal transition
**fr** fasciculus retroflexus
**LEC** lateral entorhinal cortex
**LM** lateral mammillary nucleus
**LP** lateral posterior thalamic nucleus
**LV** lateral ventricle
**mfb** medial forebrain bundle
**MG** med geniculate nucleus
**ml** medial lemniscus
**MM** med mamm nu, med part
**mp** mammillary peduncle
**MPT** medial pretectal nucleus
**opt** optic tract
**PAG** periaqueductal gray
**PaS** parasubiculum
**pc** posterior commissure
**PiRe** pineal recess
**Po** posterior thalamic nuclear group
**PRh35** perirhinal cortex, area 35
**PRh36** perirhinal cortex, area 36
**PrS** presubiculum
**RS23** retrosplenial ctx area 23
**RS29ab** retrosplenial ctx area 29ab
**RS29c** retrosplenial ctx area 29c
**RS30** retrosplenial ctx area 30
**S** subiculum
**SNC** substantia nigra, compact part
**SNR** substantia nigra, reticular part
**VC** visual cortex
**VPM** vent post med thal nucleus
**ZI** zona incerta

## Plate 67 - Sinus: 3.41 mm

Interaural: 6.11 mm
Nissl section 138

Interaural: 6.07 mm
AChE section 138

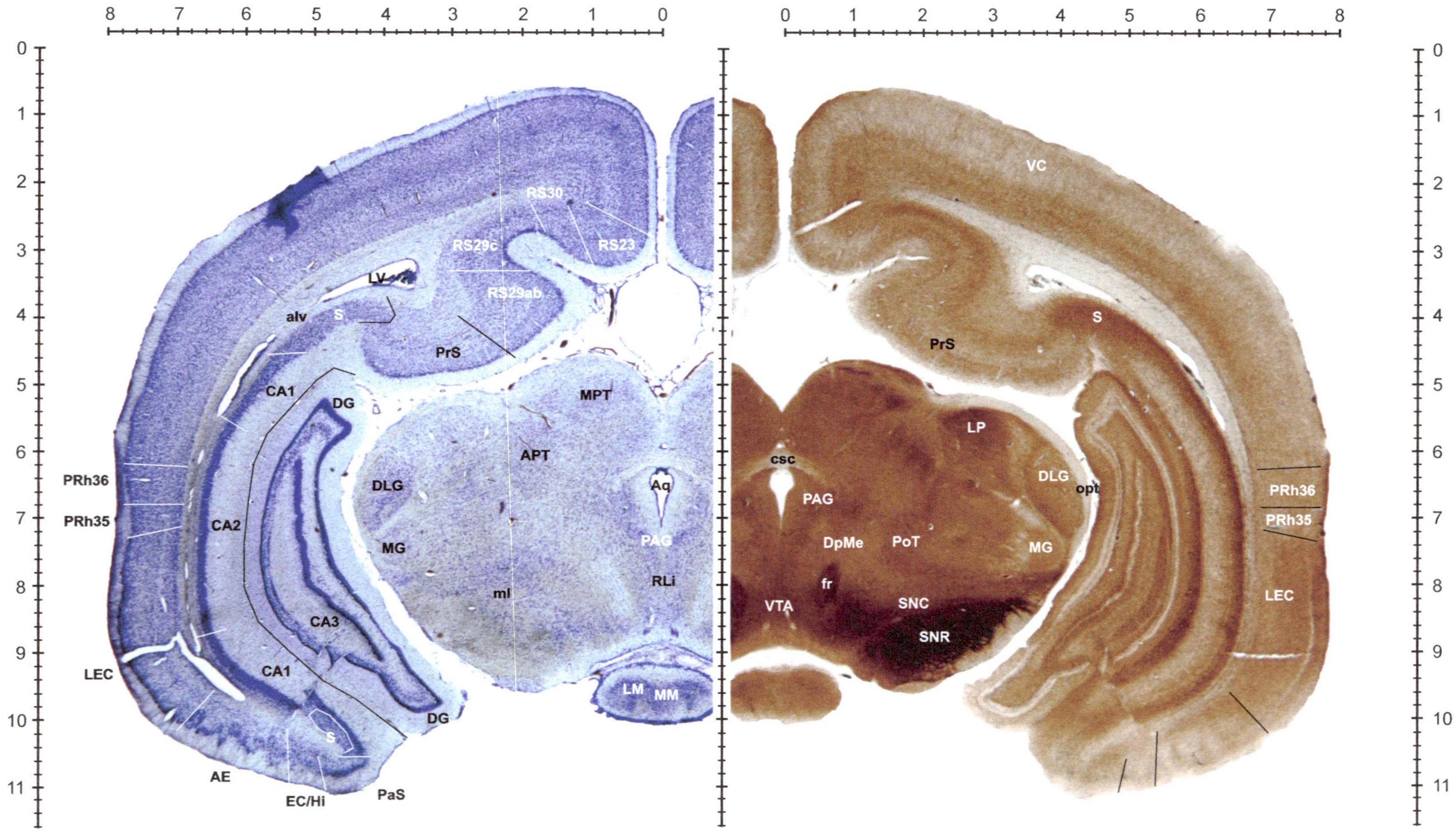


**AE** amygdalo-entorhinal transition area
**alv** alveus of the hippocampus
**APT** anterior pretectal nucleus
**Aq** aqueduct
**CA1** field CA1 of the hippocampus
**CA2** field CA2 of the hippocampus
**CA3** field CA3 of the hippocampus
**csc** commissure of the superior colliculus
**DG** dentate gyrus
**DLG** dorsal lateral geniculate nucleus
**DpMe** deep mesencephalic nucleus
**EC/Hi** entorhinal cx-hippocampal transition
**fr** fasciculus retroflexus
**LEC** lateral entorhinal cortex
**LM** lateral mammillary nucleus
**LP** lat post thal nucleus
**LV** lateral ventricle
**MG** med geniculate nucleus
**ml** medial lemniscus
**MM** med mamm nucleus, med part
**MPT** medial pretectal nucleus
**opt** optic tract
**PaS** parasubiculum
**PAG** periaqueductal gray
**PoT** posterior thalamic nu group, triangular
**PRh35** perirhinal cortex, area 35
**PRh36** perirhinal cortex, area 36
**PrS** presubiculum
**RLi** rostral linear nucleus (midbrain)
**RS23** retrosplenial ctx area 23
**RS29ab** retrosplenial ctx area 29ab
**RS29c** retrosplenial ctx area 29c
**RS30** retrosplenial ctx area 30
**S** subiculum
**SNC** substantia nigra, compact part
**SNR** substantia nigra, reticular part
**VC** visual cortex
**VTA** ventral tegmental area

**Plate 68 - Sinus: 3.25 mm**

Interaural: 5.95 mm
Nissl section 140

Interaural: 5.91 mm
Parv section 140

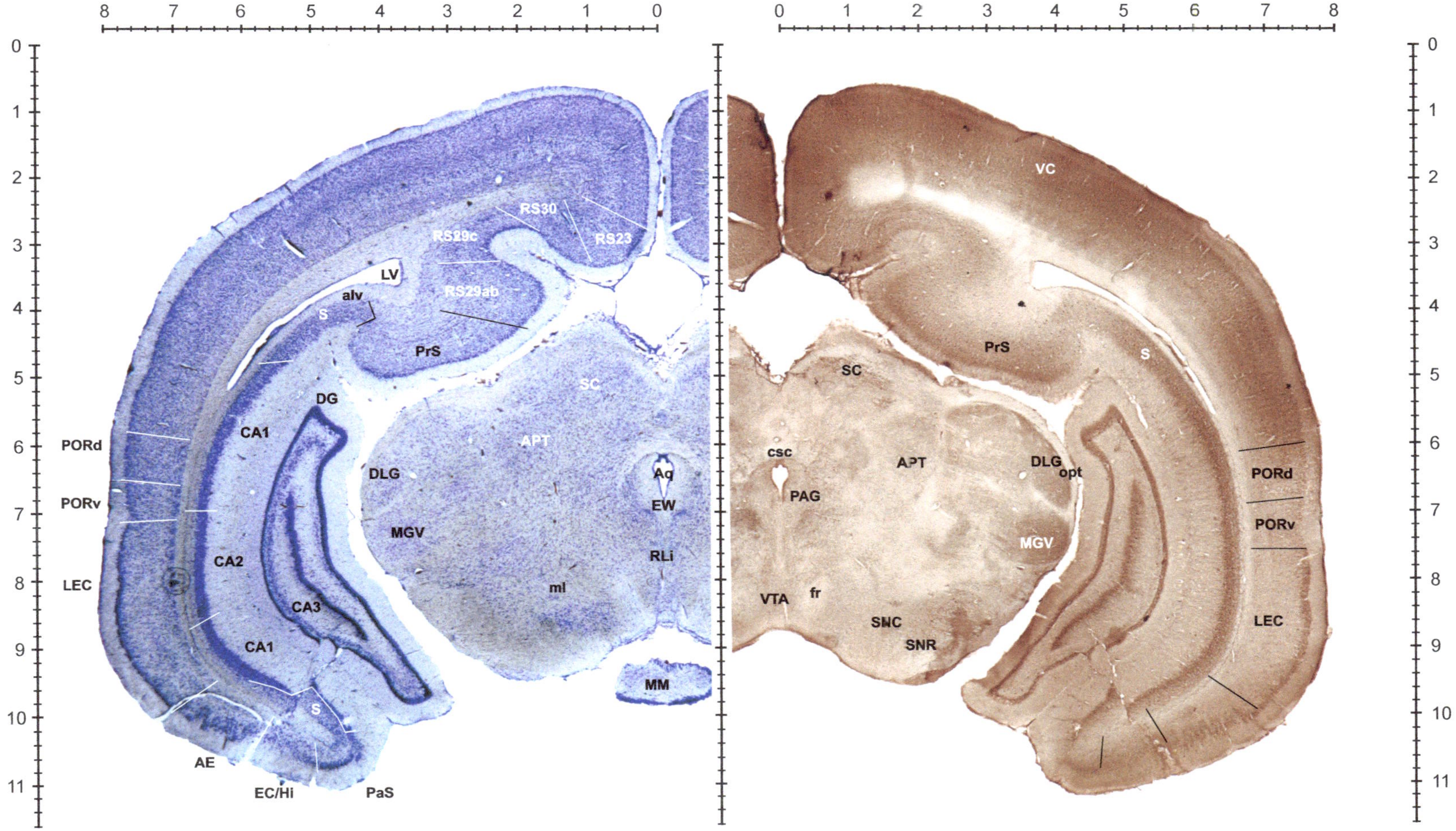


**AE** amygdalo-entorhinal transition area
**alv** alveus of the hippocampus
**APT** anterior pretectal nucleus
**Aq** aqueduct
**CA1** field CA1 of the hippocampus
**CA2** field CA2 of the hippocampus
**CA3** field CA3 of the hippocampus
**csc** commissure of the superior colliculus
**DG** dentate gyrus
**DLG** dorsal lateral geniculate nucleus
**EC/Hi** entorhinal cx-hippocampal transition
**EW** Edinger-Westphal nucleus
**fr** fasciculus retroflexus
**LEC** lateral entorhinal cortex
**LV** lateral ventricle
**MGV** med geniculate nucleus, ventral part
**ml** medial lemniscus
**MM** med mamm nucleus, med part
**opt** optic tract
**PaS** parasubiculum
**PAG** periaqueductal gray
**PORd** postrhinal cortex, dorsal
**PORv** postrhinal cortex, ventral
**PrS** presubiculum
**RLi** rostral linear nucleus (midbrain)
**RS23** retrosplenial ctx area 23
**RS29ab** retrosplenial ctx area 29ab
**RS29c** retrosplenial ctx area 29c
**RS30** retrosplenial ctx area 30
**S** subiculum
**SC** superior colliculus
**SNC** substantia nigra, compact part
**SNR** substantia nigra, reticular part
**VC** visual cortex
**VTA** ventral tegmental area

## Plate 69 - Sinus: 3.08 mm

Interaural: 5.78 mm
Nissl section 142

Interaural: 5.74 mm
AChE section 142

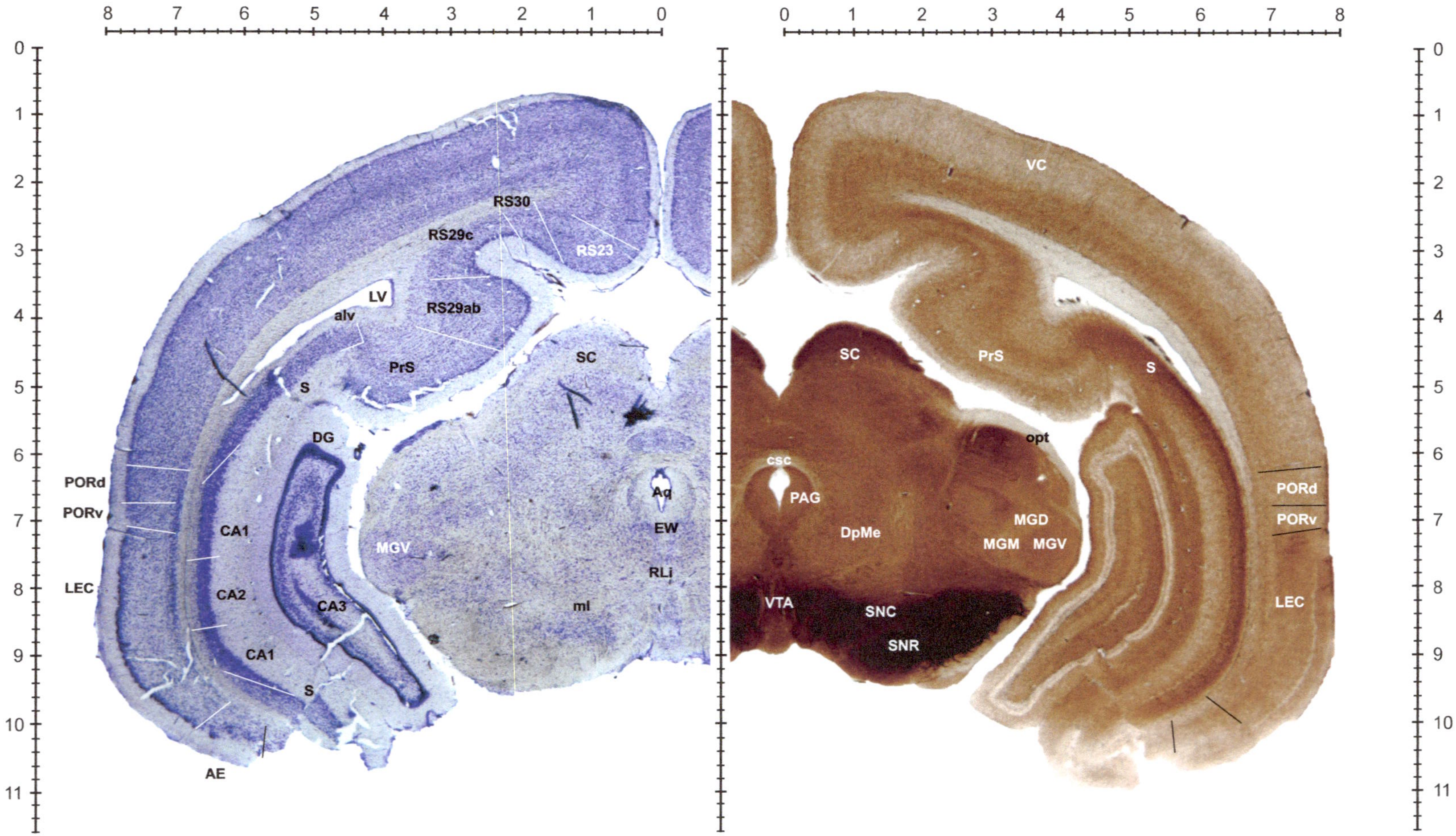


**AE** amygdalo-entorhinal transition area
**alv** alveus of the hippocampus
**Aq** aqueduct
**CA1** field CA1 of the hippocampus
**CA2** field CA2 of the hippocampus
**CA3** field CA3 of the hippocampus
**csc** commissure of the superior colliculus
**DG** dentate gyrus
**DpMe** deep mesencephalic nucleus
**EW** Edinger-Westphal nucleus
**LEC** lateral entorhinal cortex
**LV** lateral ventricle
**MGD** med geniculate nucleus, dorsal part
**MGM** med geniculate nucleus, medial part
**MGV** med geniculate nucleus, ventral part
**ml** medial lemniscus
**opt** optic tract
**PAG** periaqueductal gray
**PORd** postrhinal cortex, dorsal
**PORv** postrhinal cortex, ventral
**PrS** presubiculum
**RLi** rostral linear nucleus (midbrain)
**RS23** retrosplenial ctx area 23
**RS29ab** retrosplenial ctx area 29ab
**RS29c** retrosplenial ctx area 29c
**RS30** retrosplenial ctx area 30
**S** subiculum
**SC** superior colliculus
**SNC** substantia nigra, compact part
**SNR** substantia nigra, reticular part
**VC** visual cortex
**VTA** ventral tegmental area

## Plate 70 - Sinus: 2.92 mm

Interaural: 5.62 mm
Nissl section 144

Interaural: 5.58 mm
AChE section 144

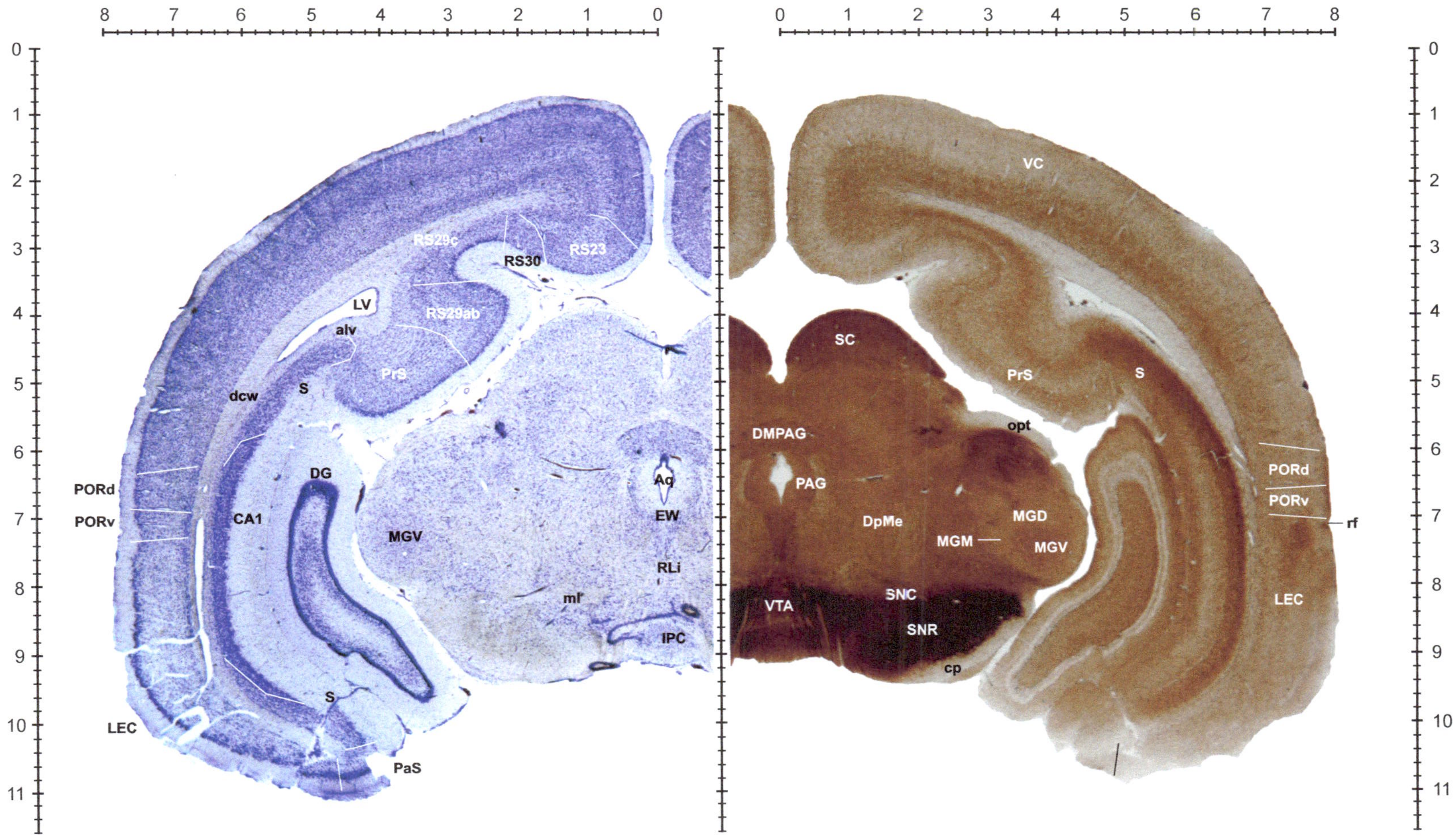


**alv** alveus of the hippocampus
**Aq** aqueduct
**cp** cerebral peduncle
**CA1** field CA1 of the hippocampus
**dcw** deep cerebral white matter
**DG** dentate gyrus
**DMPAG** dorsomedial periaqueductal gray
**DpMe** deep mesencephalic nucleus
**EW** Edinger-Westphal nucleus
**IPC** interpeduncular nu, caudal
**LEC** lateral entorhinal cortex
**LV** lateral ventricle
**MGD** med geniculate nucleus, dorsal part
**MGM** med geniculate nucleus, medial part
**MGV** med geniculate nucleus, ventral part
**ml** medial lemniscus
**opt** optic tract
**PAG** periaqueductal gray
**PaS** parasubiculum
**PORd** postrhinal cortex, dorsal
**PORv** postrhinal cortex, ventral
**PrS** presubiculum
**rf** rhinal fissure
**RLi** rostral linear nucleus (midbrain)
**RS23** retrosplenial ctx area 23
**RS29ab** retrosplenial ctx area 29ab
**RS29c** retrosplenial ctx area 29c
**RS30** retrosplenial ctx area 30
**S** subiculum
**SC** superior colliculus
**SNC** substantia nigra, compact part
**SNR** substantia nigra, reticular part
**VC** visual cortex
**VTA** ventral tegmental area

## Plate 71 - Sinus: 2.76 mm

Interaural: 5.46 mm
Nissl section 146

Interaural: 5.41 mm
Parv section 146

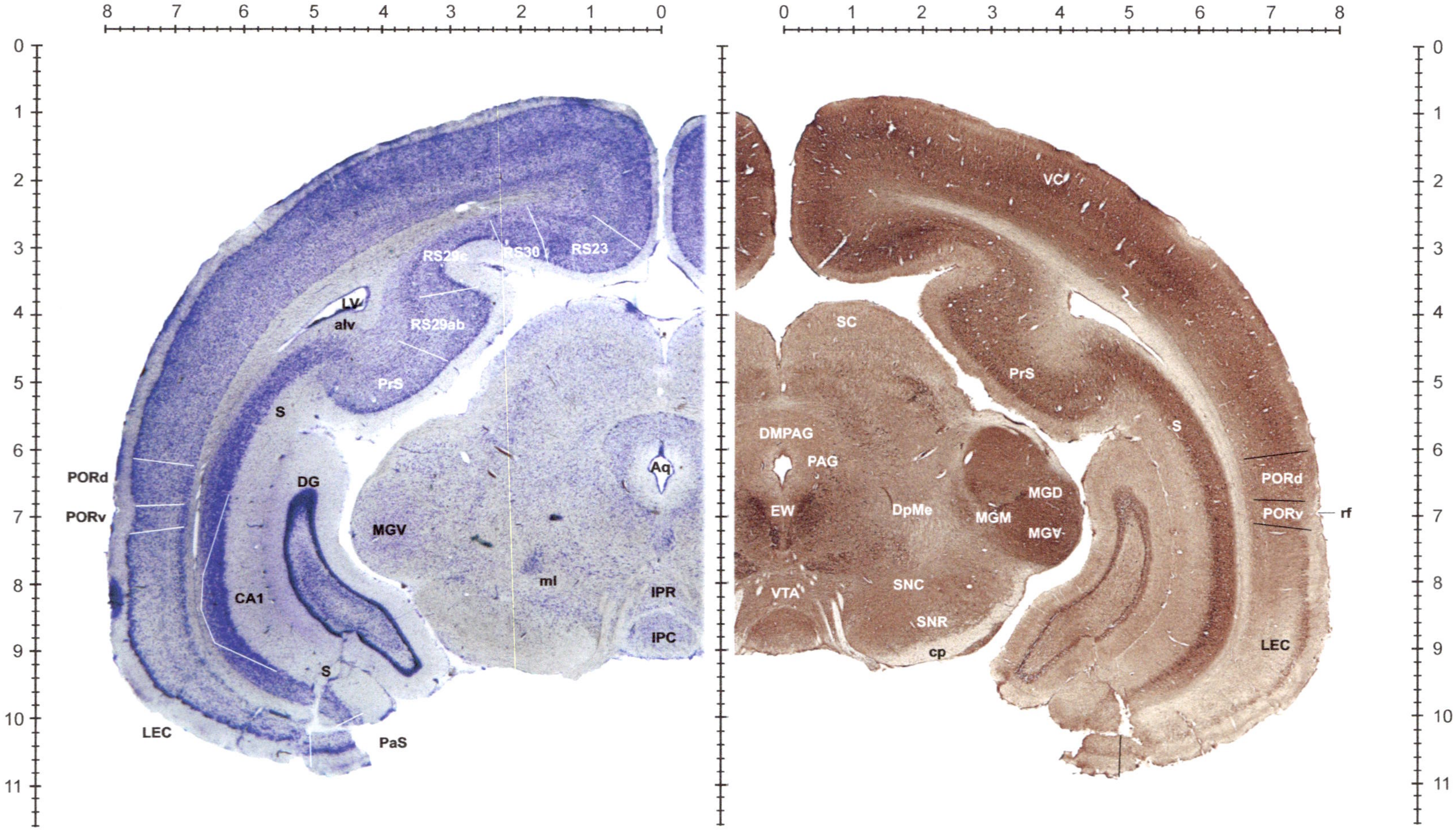


**alv** alveus of the hippocampus
**Aq** aqueduct
**cp** cerebral peduncle
**CA1** field CA1 of the hippocampus
**DG** dentate gyrus
**DMPAG** dorsomedial periaqueductal gray
**DpMe** deep mesencephalic nucleus
**EW** Edinger-Westphal nucleus
**IPC** interpeduncular nucleus, caudal
**IPR** interpeduncular nucleus, rostral
**LEC** lateral entorhinal cortex
**LV** lateral ventricle
**MGD** med geniculate nucleus, dorsal part
**MGM** med geniculate nucleus, medial part
**MGV** med geniculate nucleus, ventral part
**ml** medial lemniscus
**PAG** periaqueductal gray
**PaS** parasubiculum
**PORd** postrhinal cortex, dorsal
**PORv** postrhinal cortex, ventral
**PrS** presubiculum
**rf** rhinal fissure
**RS23** retrosplenial ctx area 23
**RS29ab** retrosplenial ctx area 29ab
**RS29c** retrosplenial ctx area 29c
**RS30** retrosplenial ctx area 30
**S** subiculum
**SC** superior colliculus
**SNC** substantia nigra, compact part
**SNR** substantia nigra, reticular part
**VC** visual cortex
**VTA** ventral tegmental area

## Plate 72 - Sinus: 2.59 mm

Interaural: 5.21 mm
Nissl section 149

Interaural: 5.25 mm
AChE section 148

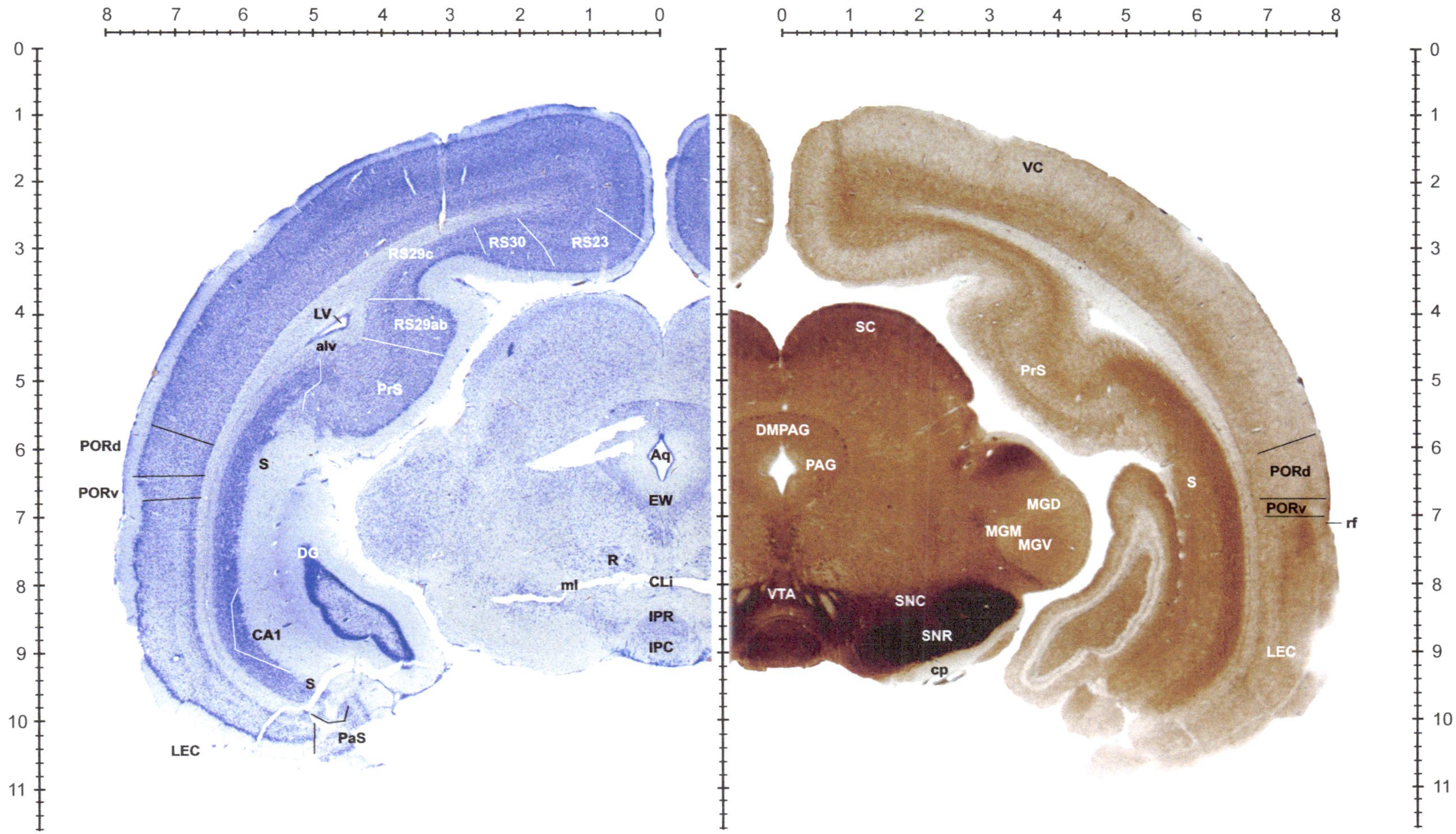


**alv** alveus of the hippocampus
**Aq** aqueduct
**cp** cerebral peduncle
**CA1** field CA1 of the hippocampus
**CLi** caudal linear nucleus of the raphe
**DG** dentate gyrus
**DMPAG** dorsomedial periaqueductal gray
**EW** Edinger-Westphal nucleus
**IPC** interpeduncular nucleus, caudal
**IPR** interpeduncular nucleus, rostral
**LEC** lateral entorhinal cortex
**LV** lateral ventricle
**MGD** med geniculate nucleus, dorsal part
**MGM** med geniculate nucleus, medial part
**MGV** med geniculate nucleus, ventral part
**ml** medial lemniscus
**PAG** periaqueductal gray
**PaS** parasubiculum
**PORd** postrhinal cortex, dorsal
**PORv** postrhinal cortex, ventral
**PrS** presubiculum
**R** red nucleus
**rf** rhinal fissure
**RS23** retrosplenial ctx area 23
**RS29ab** retrosplenial ctx area 29ab
**RS29c** retrosplenial ctx area 29c
**RS30** retrosplenial ctx area 30
**S** subiculum
**SC** superior colliculus
**SNC** substantia nigra, compact part
**SNR** substantia nigra, reticular part
**VC** visual cortex
**VTA** ventral tegmental area

## Plate 73 - Sinus: 2.43 mm

Interaural: 5.13 mm
Nissl section 150

Interaural: 5.09 mm
AChE section 150

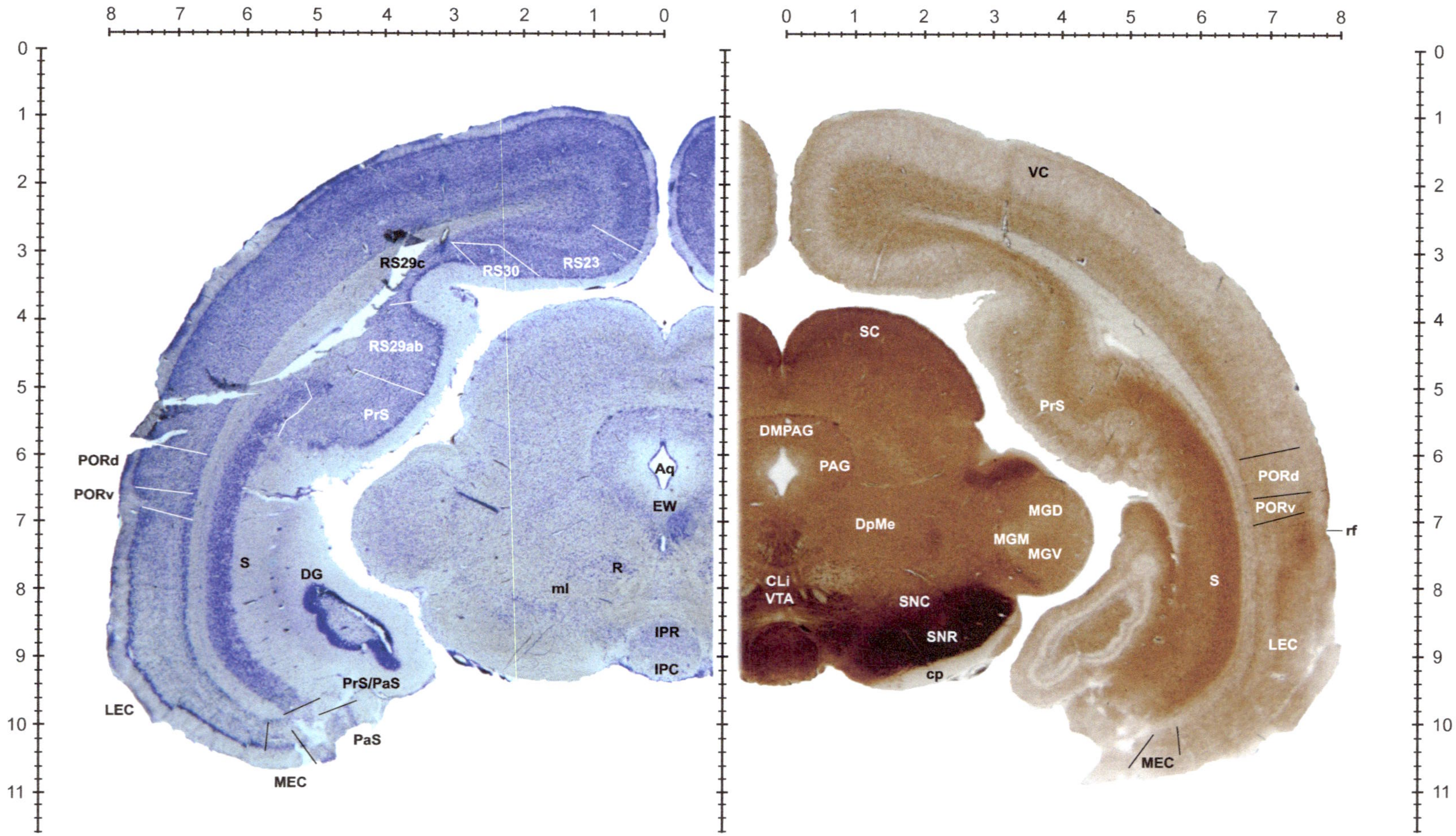


- **Aq** aqueduct
- **cp** cerebral peduncle
- **CLi** caudal linear nucleus of the raphe
- **DG** dentate gyrus
- **DMPAG** dorsomedial periaqueductal gray
- **DpMe** deep mesencephalic nucleus
- **EW** Edinger-Westphal nucleus
- **IPC** interpeduncular nucleus, caudal
- **IPR** interpeduncular nucleus, rostral
- **LEC** lateral entorhinal cortex
- **MEC** medial entorhinal cortex
- **MGD** med geniculate nucleus, dorsal part
- **MGM** med geniculate nucleus, medial part
- **MGV** med geniculate nucleus, ventral part
- **ml** medial lemniscus
- **PAG** periaqueductal gray
- **PaS** parasubiculum
- **PORd** postrhinal cortex, dorsal
- **PORv** postrhinal cortex, ventral
- **PrS** presubiculum
- **R** red nucleus
- **rf** rhinal fissure
- **RS23** retrosplenial ctx area 23
- **RS29ab** retrosplenial ctx area 29ab
- **RS29c** retrosplenial ctx area 29c
- **RS30** retrosplenial ctx area 30
- **S** subiculum
- **SC** superior colliculus
- **SNC** substantia nigra, compact part
- **SNR** substantia nigra, reticular part
- **VC** visual cortex
- **VTA** ventral tegmental area

## Plate 74 - Sinus: 2.26 mm

Interaural: 4.96 mm
N ssl section 152

Interaural: 4.92 mm
Parv section 152

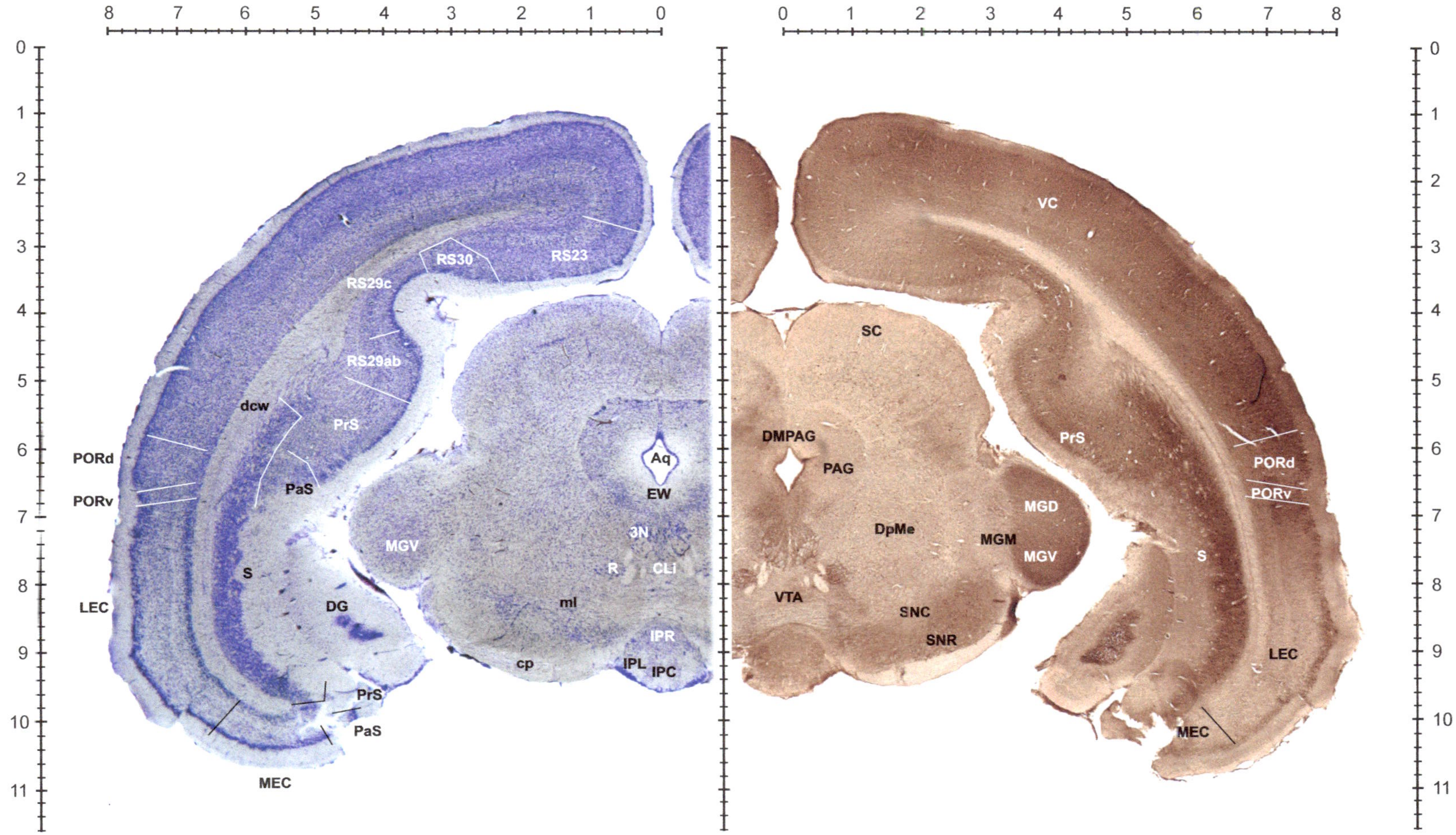


**3N** oculomotor nucleus
**Aq** aqueduct
**cp** cerebral peduncle
**CLi** caudal linear nucleus of the raphe
**dcw** deep cerebral white matter
**DG** dentate gyrus
**DMPAG** dorsomedial periaqueductal gray
**DpMe** deep mesencephalic nucleus
**EW** Edinger-Westphal nucleus
**IPC** interpeduncular nucleus, caudal
**IPL** interpeduncular nucleus, lateral
**IPR** interpeduncular nucleus, rostral
**LEC** lateral entorhinal cortex
**MEC** medial entorhinal cortex
**MGD** med geniculate nucleus, dorsal part
**MGM** med geniculate nucleus, medial part
**MGV** med geniculate nucleus, ventral part
**ml** medial lemniscus
**PAG** periaqueductal gray
**PaS** parasubiculum
**PORd** postrhinal cortex, dorsal
**PORv** postrhinal cortex, ventral
**PrS** presubiculum
**R** red nucleus
**RS23** retrosplenial ctx area 23
**RS29ab** retrosplenial ctx area 29ab
**RS29c** retrosplenial ctx area 29c
**RS30** retrosplenial ctx area 30
**S** subiculum
**SC** superior colliculus
**SNC** substantia nigra, compact part
**SNR** substantia nigra, reticular part
**VC** visual cortex
**VTA** ventral tegmental area

## Plate 75 - Sinus: 2.02 mm

Interaural: 4.72 mm
Nissl section 155

Interaural: 4.68 mm
Parv section 155

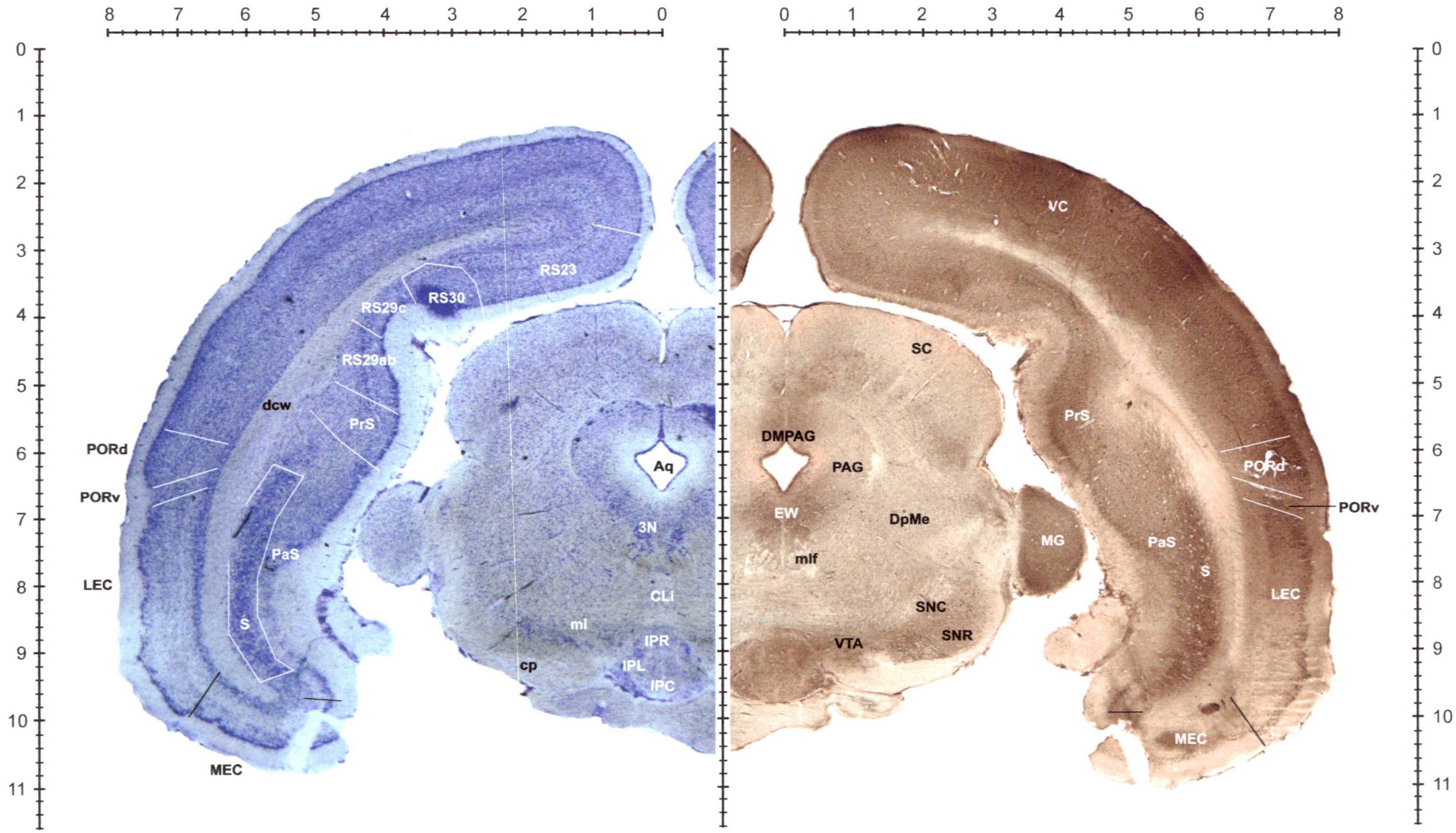


**3N** oculomotor nucleus
**Aq** aqueduct
**cp** cerebral peduncle
**CLi** caudal linear nucleus of the raphe
**dcw** deep cerebral white matter
**DMPAG** dorsomedial periaqueductal gray
**DpMe** deep mesencephalic nucleus
**EW** Edinger-Westphal nucleus
**IPC** interpeduncular nucleus, caudal
**IPL** interpeduncular nucleus, lateral
**IPR** interpeduncular nucleus, rostral
**LEC** lateral entorhinal cortex
**MEC** medial entorhinal cortex
**MG** med geniculate nucleus
**ml** medial lemniscus
**mlf** medial longitudinal fasciculus
**PAG** periaqueductal gray
**PaS** parasubiculum
**PORd** postrhinal cortex, dorsal
**PORv** postrhinal cortex, ventral
**PrS** presubiculum
**RS23** retrosplenial ctx area 23
**RS29ab** retrosplenial ctx area 29ab
**RS29c** retrosplenial ctx area 29c
**RS30** retrosplenial ctx area 30
**S** subiculum
**SC** superior colliculus
**SNC** substantia nigra, compact part
**SNR** substantia nigra, reticular part
**VC** visual cortex
**VTA** ventral tegmental area

## Plate 76 - Sinus: 1.85 mm

Interaural: 4.55 mm
Nissl section 157

Interaural: 4.51 mm
AChE section 157

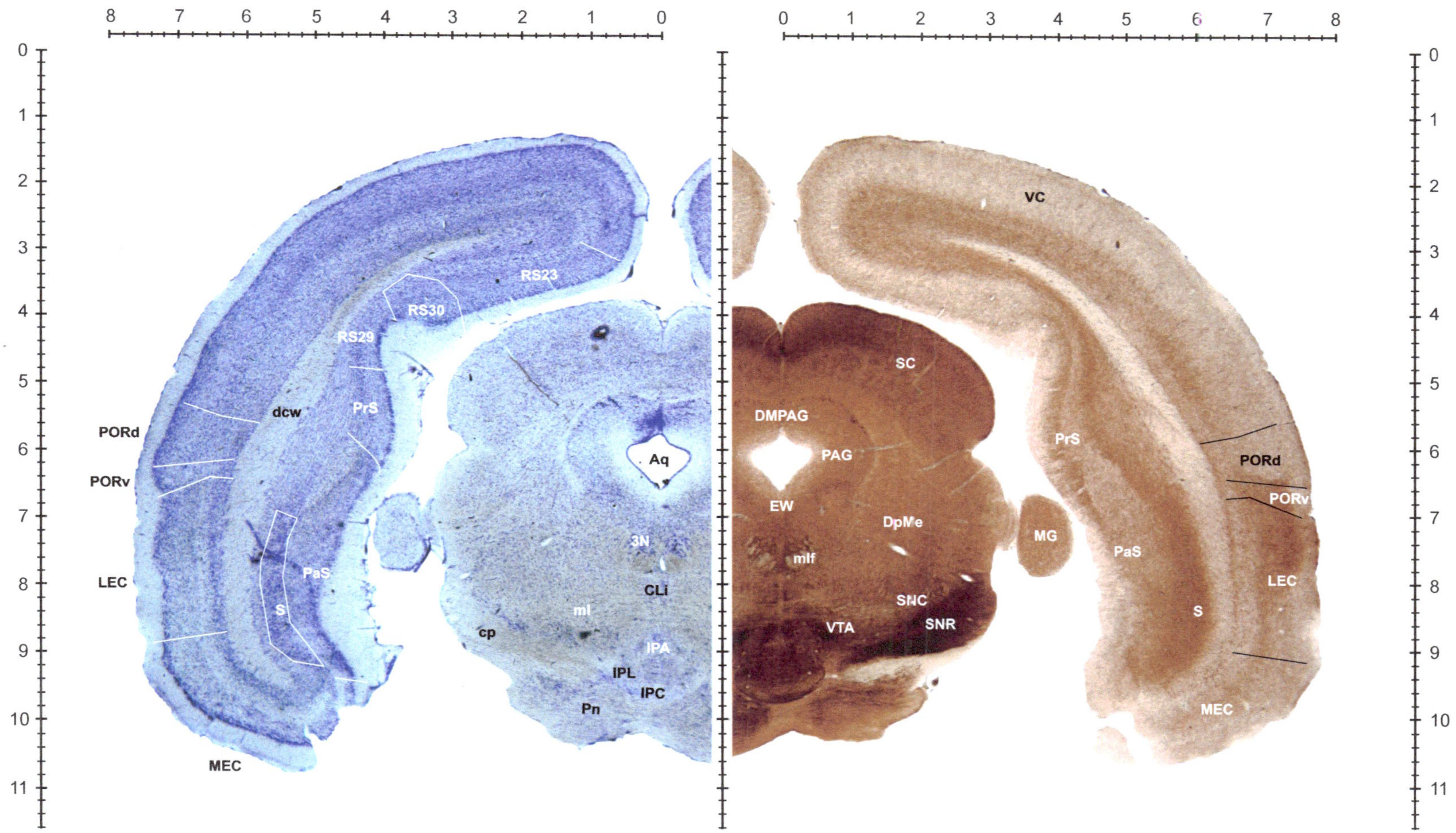


**3N** oculomotor nucleus
**Aq** aqueduct
**cp** cerebral peduncle
**CLi** caudal linear nucleus of the raphe
**dcw** deep cerebral white matter
**DMPAG** dorsomedial periaqueductal gray
**DpMe** deep mesencephalic nucleus
**EW** Edinger-Westphal nucleus
**IPA** interpeduncular nucleus, apical
**IPC** interpeduncular nucleus, caudal
**IPL** interpeduncular nucleus, lateral
**LEC** lateral entorhinal cortex
**MEC** medial entorhinal cortex
**MG** med geniculate nucleus
**ml** medial lemniscus
**mlf** medial longitudinal fasciculus
**PAG** periaqueductal gray
**PaS** parasubiculum
**Pn** pontine nuclei
**PORd** postrhinal cortex, dorsal
**PORv** postrhinal cortex, ventral
**PrS** presubiculum
**RS23** retrosplenial ctx area 23
**RS29** retrosplenial ctx area 29
**RS30** retrosplenial ctx area 30
**S** subiculum
**SC** superior colliculus
**SNC** substantia nigra, compact part
**SNR** substantia nigra, reticular part
**VC** visual cortex
**VTA** ventral tegmental area

**Plate 77 - Sinus: 1.77 mm**

Interaural: 4.47 mm
Nissl section 158

Interaural: 4.43 mm
Parv section 158

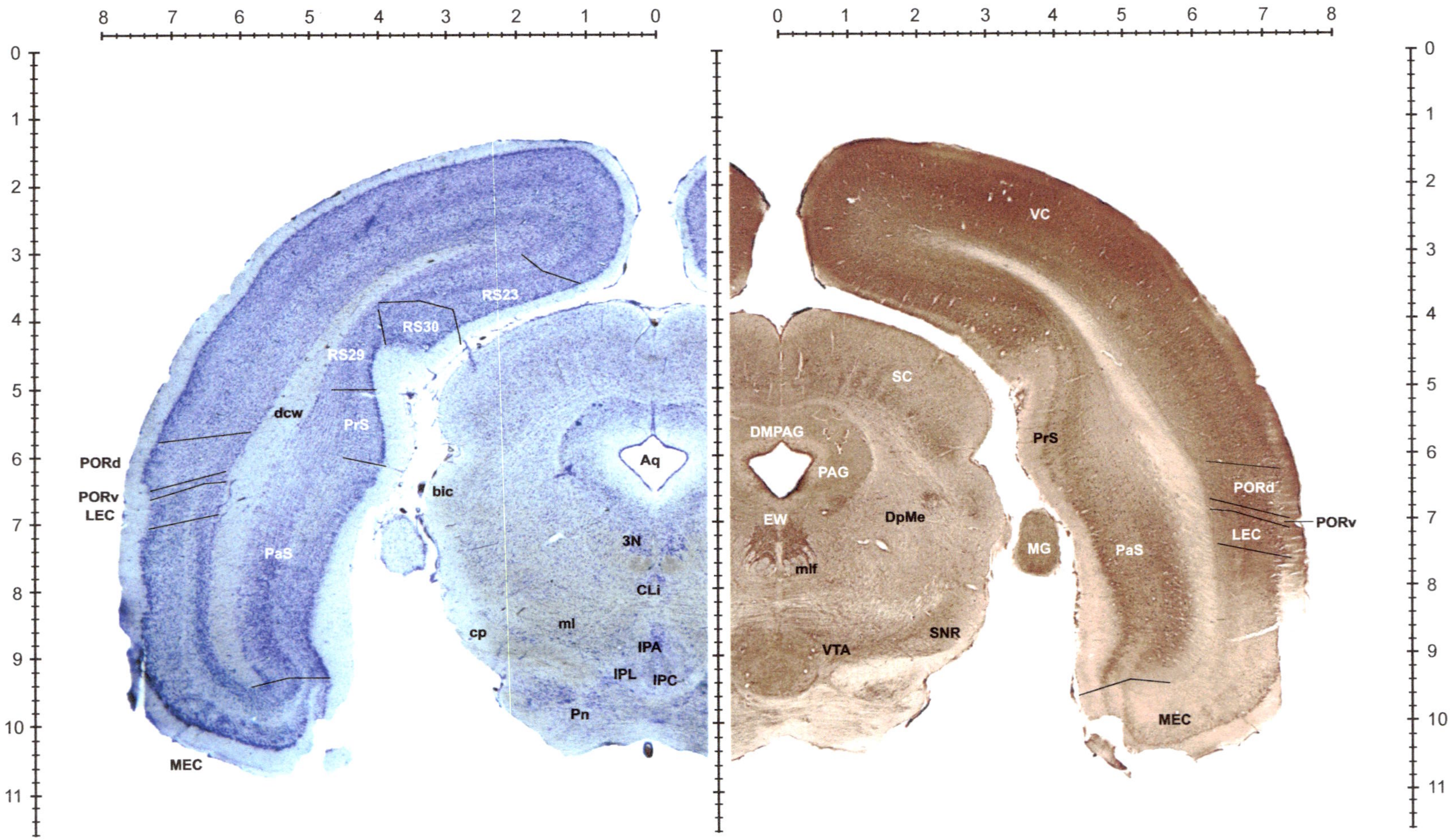


**3N** oculomotor nucleus
**Aq** aqueduct
**bic** brachium of the inferior colliculus
**cp** cerebral peduncle
**CLi** caudal linear nucleus of the raphe
**dcw** deep cerebral white matter
**DMPAG** dorsomedial periaqueductal gray
**DpMe** deep mesencephalic nucleus
**EW** Edinger-Westphal nucleus
**IPA** interpeduncular nucleus, apical
**IPC** interpeduncular nucleus, caudal
**IPL** interpeduncular nucleus, lateral
**LEC** lateral entorhinal cortex
**MEC** medial entorhinal cortex
**MG** med geniculate nucleus
**ml** medial lemniscus
**mlf** medial longitudinal fasciculus
**PAG** periaqueductal gray
**PaS** parasubiculum
**Pn** pontine nuclei
**PORd** postrhinal cortex, dorsal
**PORv** postrhinal cortex, ventral
**PrS** presubiculum
**RS23** retrosplenial ctx area 23
**RS29** retrosplenial ctx area 29
**RS30** retrosplenial ctx area 30
**SC** superior colliculus
**SNR** substantia nigra, reticular part
**VC** visual cortex
**VTA** ventral tegmental area

**Plate 78 - Sinus: 1.61 mm**

Interaural: 4.31 mm
Nissl section 160

Interaural: 4.27 mm
AChE section 160

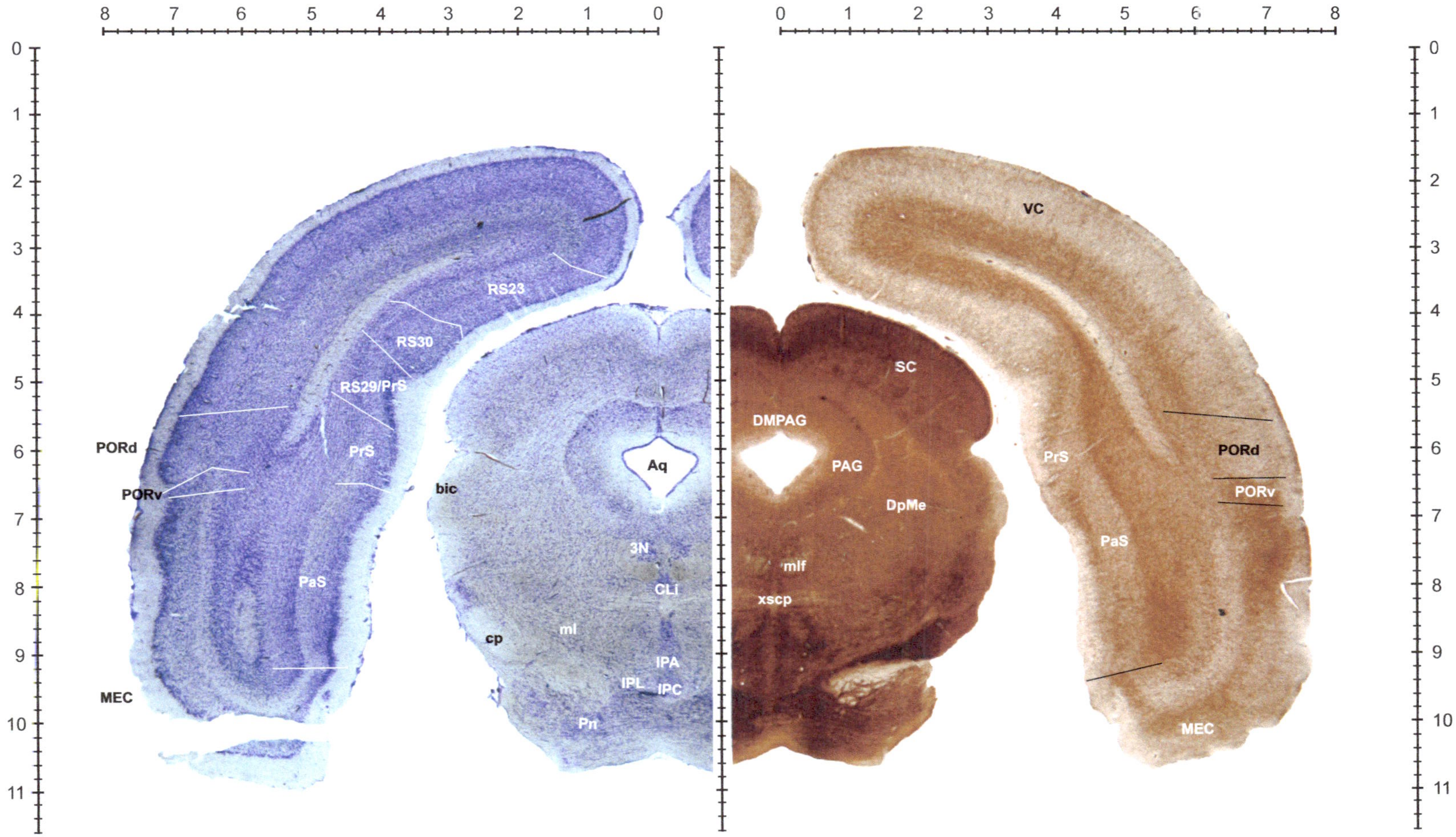


**3N** oculomotor nucleus
**Aq** aqueduct
**bic** brachium of the inferior colliculus
**cp** cerebral peduncle
**CLi** caudal linear nucleus of the raphe
**DMPAG** dorsomedial periaqueductal gray
**DpMe** deep mesencephalic nucleus
**IPA** interpeduncular nucleus, apical
**IPC** interpeduncular nucleus, caudal
**IPL** interpeduncular nucleus, lateral
**MEC** medial entorhinal cortex
**ml** medial lemniscus
**mlf** medial longitudinal fasciculus
**PAG** periaqueductal gray
**PaS** parasubiculum
**Pn** pontine nuclei
**PORd** postrhinal cortex, dorsal
**PORv** postrhinal cortex, ventral
**PrS** presubiculum
**RS23** retrosplenial ctx area 23
**RS29** retrosplenial ctx area 29
**RS30** retrosplenial ctx area 30
**SC** superior colliculus
**VC** visual cortex
**xscp** decuss sup cereb peduncle

## Plate 79 - Sinus: 1.53 mm

Interaural: 4.23 mm
Nissl section 161

Interaural: 4.18 mm
Parv section 161

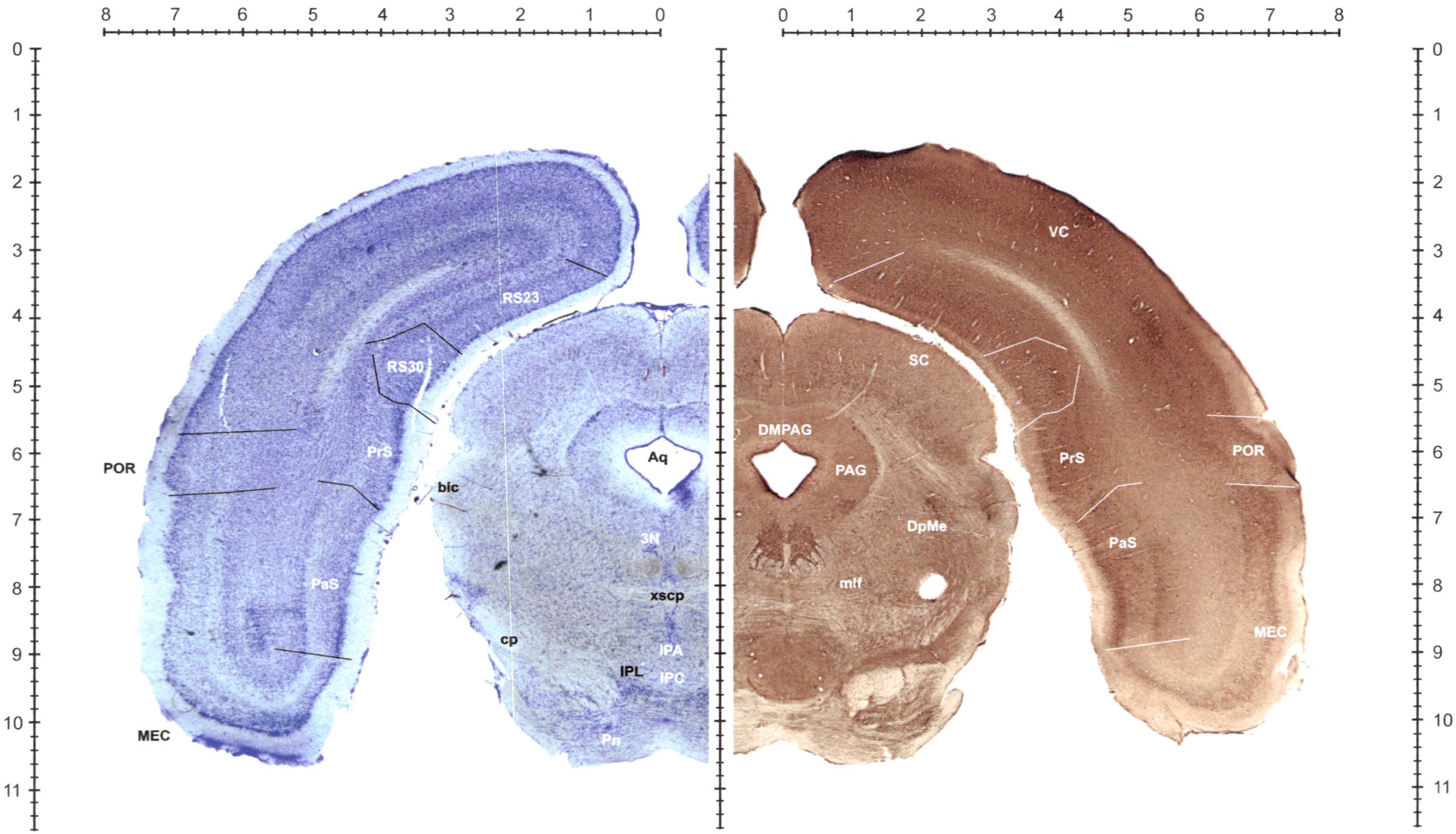


**3N** oculomotor nucleus
**Aq** aqueduct
**bic** brachium of the inferior colliculus
**cp** cerebral peduncle
**DMPAG** dorsomedial periaqueductal gray
**DpMe** deep mesencephalic nucleus
**IPA** interpeduncular nucleus, apical
**IPC** interpeduncular nucleus, caudal
**IPL** interpeduncular nucleus, lateral
**MEC** medial entorhinal cortex
**mlf** medial longitudinal fasciculus
**PAG** periaqueductal gray
**PaS** parasubiculum
**Pn** pontine nuclei
**POR** postrhinal cortex
**PrS** presubiculum
**RS23** retrosplenial ctx area 23
**RS30** retrosplenial ctx area 30
**SC** superior colliculus
**VC** visual cortex
**xscp** decuss sup cereb peduncle

**Plate 80 - Sinus: 1.36 mm**

Interaural: 4.06 mm
Nissl section 163

Interaural: 4.02 mm
AChE section 163

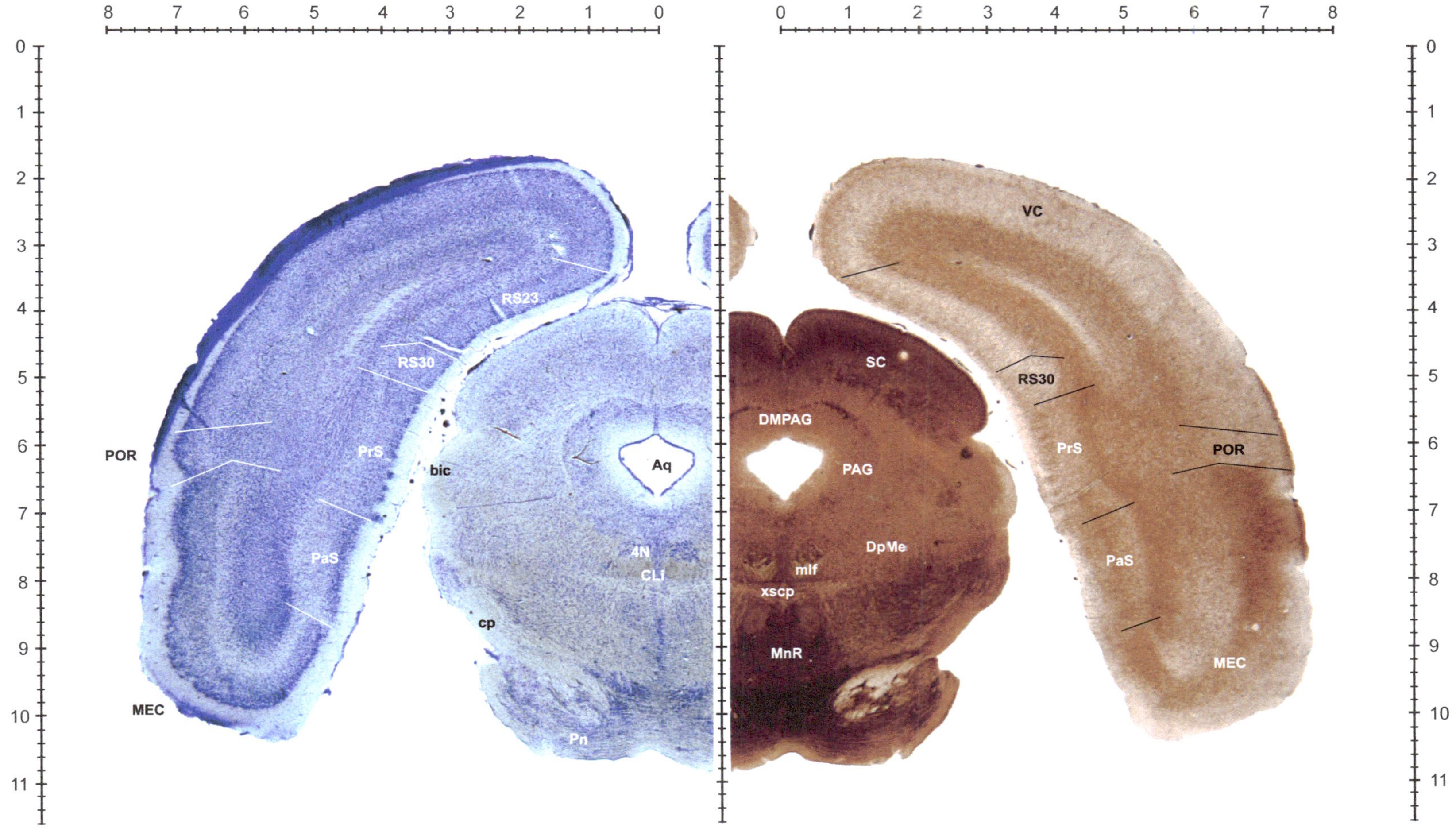


**4N** trochlear nucleus
**Aq** aqueduct
**bic** brachium of the inferior colliculus
**CLi** caudal linear nucleus of the raphe
**cp** cerebral peduncle
**DMPAG** dorsomedial periaqueductal gray
**DpMe** deep mesencephalic nucleus
**MEC** medial entorhinal cortex
**mlf** medial longitudinal fasciculus
**MnR** median raphe nucleus
**PAG** periaqueductal gray
**PaS** parasubiculum
**Pn** pontine nuclei
**POR** postrhinal cortex
**PrS** presubiculum
**RS23** retrosplenial ctx area 23
**RS30** retrosplenial ctx area 30
**SC** superior colliculus
**VC** visual cortex
**xscp** decuss sup cereb peduncle

**Plate 81 - Sinus: 1.28 mm**

Interaural: 3.98 mm
Nissl section 164

Interaural: 3.94 mm
Parv section 164

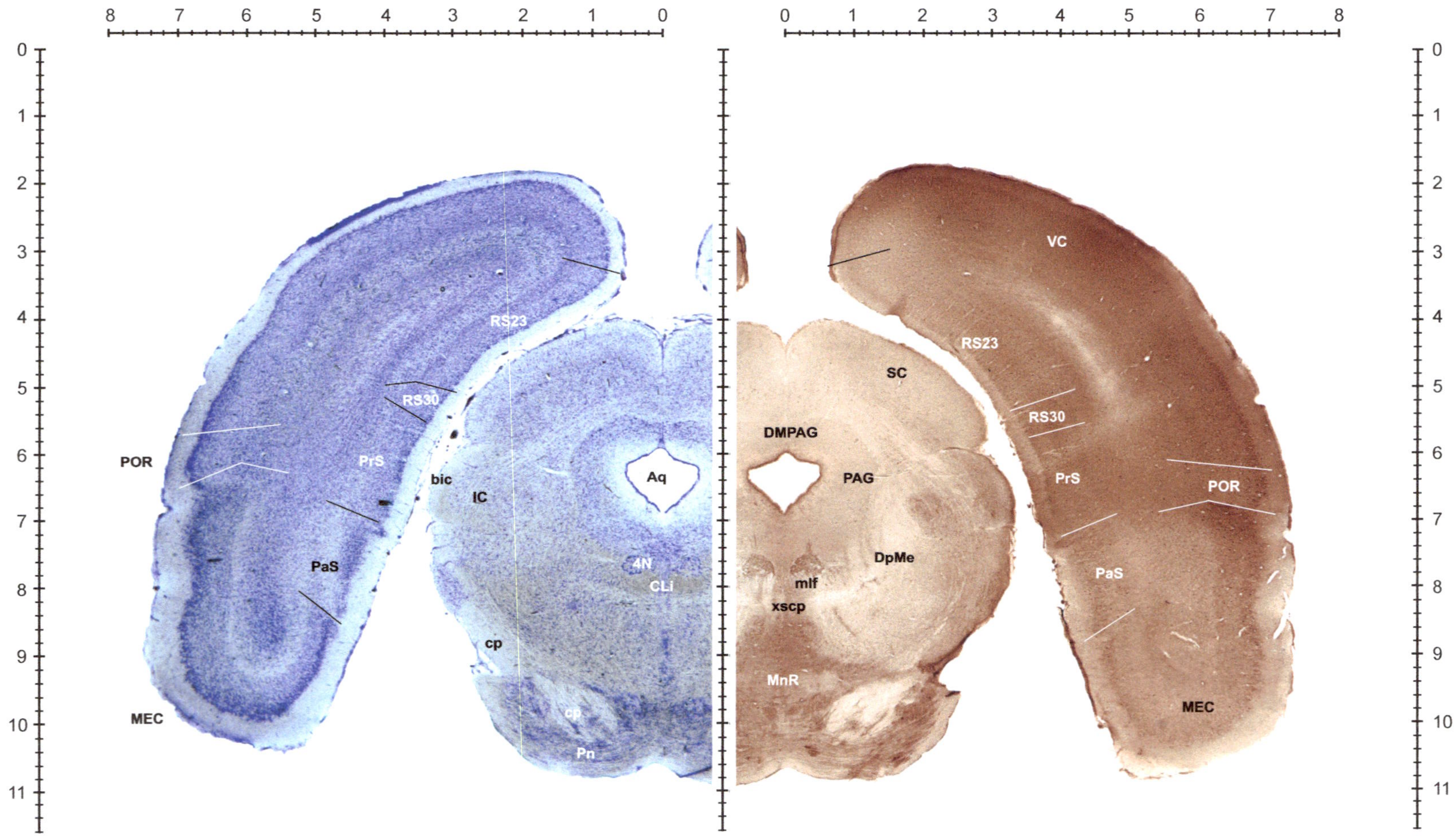


**4N** trochlear nucleus
**Aq** aqueduct
**bic** brachium of the inferior colliculus
**CLi** caudal linear nucleus of the raphe
**cp** cerebral peduncle
**DMPAG** dorsomedial periaqueductal gray
**DpMe** deep mesencephalic nucleus
**IC** inferior colliculus
**MEC** medial entorhinal cortex
**mlf** medial longitudinal fasciculus
**MnR** median raphe nucleus
**PAG** periaqueductal gray
**PaS** parasubiculum
**Pn** pontine nuclei
**POR** postrhinal cortex
**PrS** presubiculum
**RS23** retrosplenial ctx area 23
**RS30** retrosplenial ctx area 30
**SC** superior colliculus
**VC** visual cortex
**xscp** decuss sup cereb peduncle

## Plate 82 - Sinus: 1.20 mm

Interaural: 3.90 mm
Nissl section 165

Interaural: 3.86 mm
AChE section 165

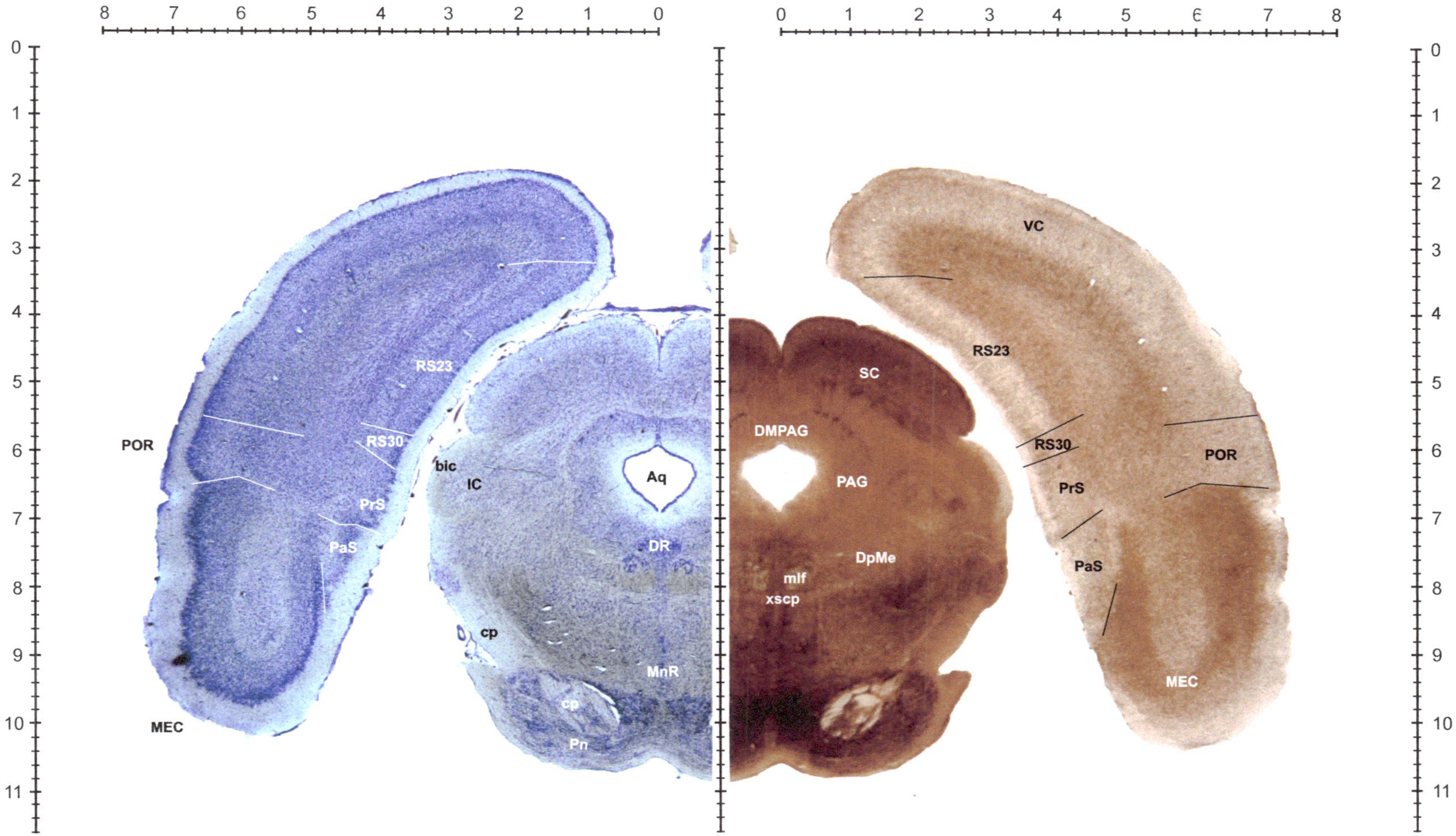


**Aq** aqueduct
**bic** brachium of the inferior colliculus
**cp** cerebral peduncle
**DMPAG** dorsomedial periaqueductal gray
**DpMe** deep mesencephalic nucleus
**DR** dorsal raphe nucleus
**IC** inferior colliculus
**MEC** medial entorhinal cortex
**mlf** medial longitudinal fasciculus
**MnR** median raphe nucleus
**PAG** periaqueductal gray
**PaS** parasubiculum
**Pn** pontine nuclei
**POR** postrhinal cortex
**PrS** presubiculum
**RS23** retrosplenial ctx area 23
**RS30** retrosplenial ctx area 30
**SC** superior colliculus
**VC** visual cortex
**xscp** decuss sup cereb peduncle

## Plate 83 - Sinus: 1.12 mm

Interaural: 3.82 mm
Nissl section 166

Interaural: 3.77 mm
AChE section 166

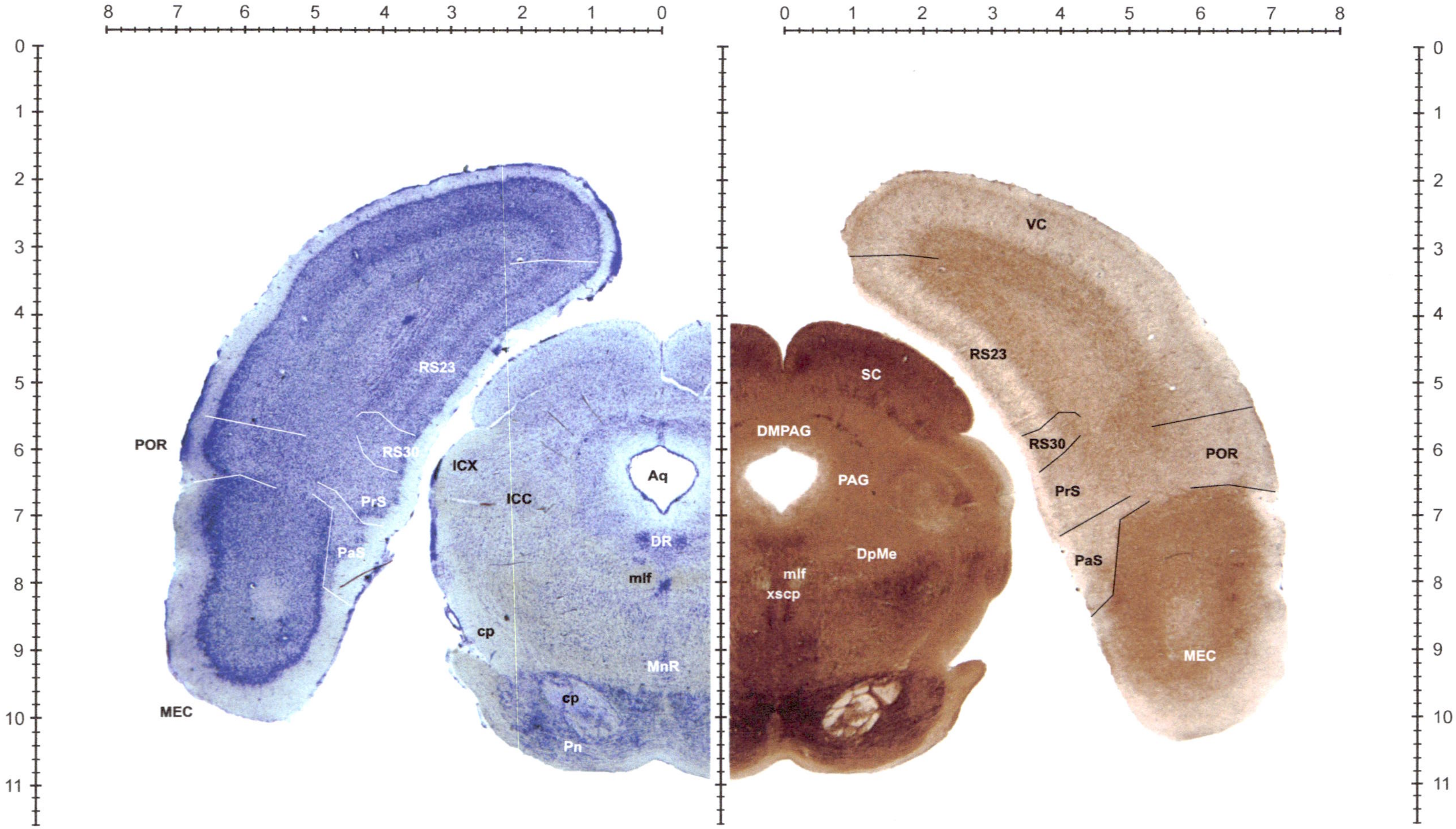


**Aq** aqueduct
**cp** cerebral peduncle
**DMPAG** dorsomedial periaqueductal gray
**DpMe** deep mesencephalic nucleus
**DR** dorsal raphe nucleus
**ICC** inferior colliculus, central nucleus
**ICX** inferior colliculus, external nucleus
**MEC** medial entorhinal cortex
**mlf** medial longitudinal fasciculus
**MnR** median raphe nucleus
**PAG** periaqueductal gray
**PaS** parasubiculum
**Pn** pontine nuclei
**POR** postrhinal cortex
**PrS** presubiculum
**RS23** retrosplenial ctx area 23
**RS30** retrosplenial ctx area 30
**SC** superior colliculus
**VC** visual cortex
**xscp** decuss sup cereb peduncle

## Plate 84 - Sinus: 1.03 mm

Interaural: 3.73 mm
Nissl section 167

Interaural: 3.69 mm
Parv section 167

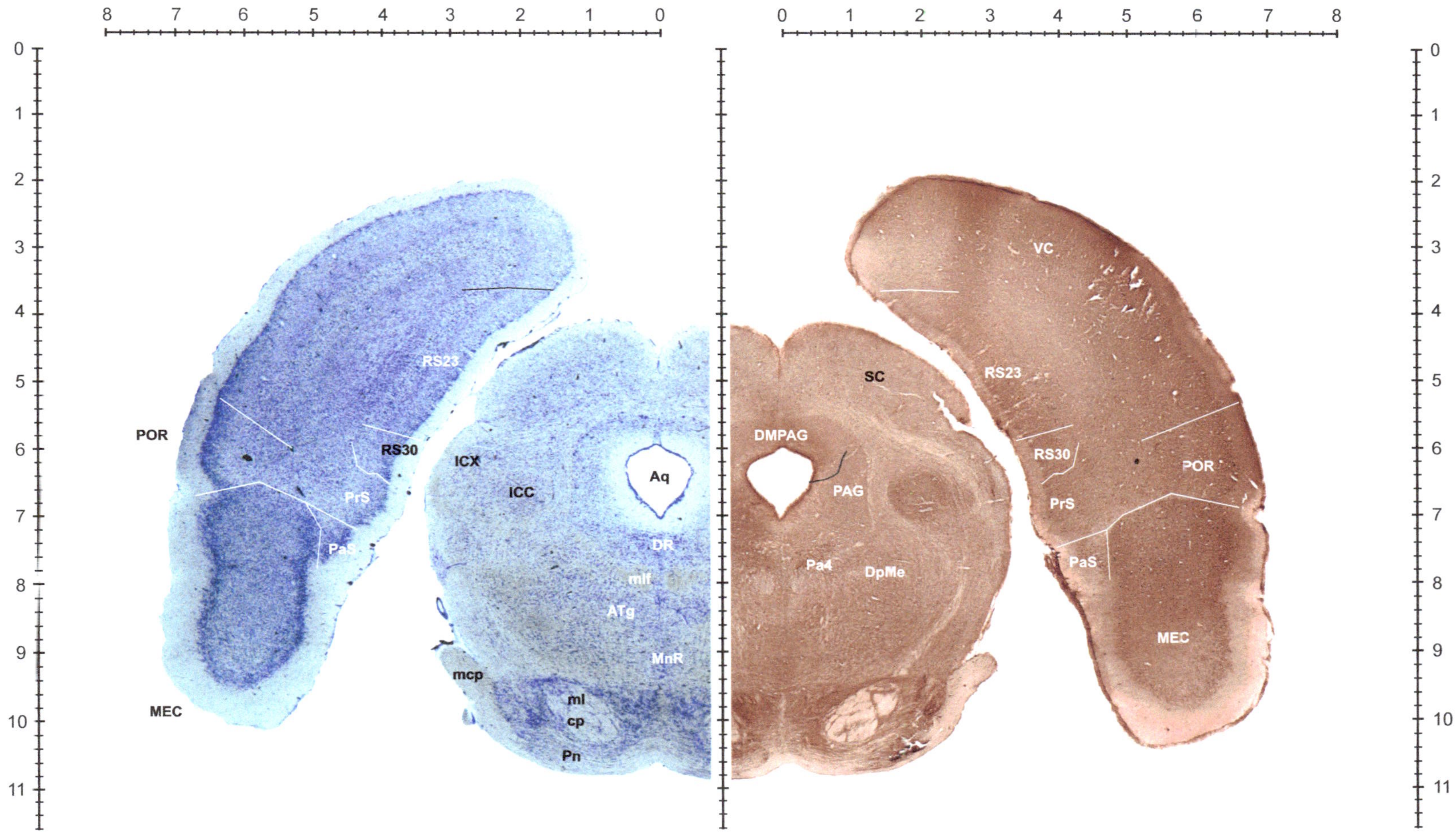


**Aq** aqueduct
**ATg** anterior tegmental nucleus
**cp** cerebral peduncle
**DMPAG** dorsomedial periaqueductal gray
**DpMe** deep mesencephalic nucleus
**DR** dorsal raphe nucleus
**ICC** inferior colliculus, central nucleus
**ICX** inferior colliculus, external nucleus
**mcp** middle cerebellar peduncle
**MEC** medial entorhinal cortex
**ml** medial lemniscus
**mlf** medial longitudinal fasciculus
**MnR** median raphe nucleus
**Pa4** paratrochlear nucleus
**PAG** periaqueductal gray
**PaS** parasubiculum
**Pn** pontine nuclei
**POR** postrhinal cortex
**PrS** presubiculum
**RS23** retrosplenial ctx area 23
**RS30** retrosplenial ctx area 30
**SC** superior colliculus
**VC** visual cortex

## Plate 85 - Sinus: 0.95 mm

Interaural: 3.65 mm
Nissl section 168

Interaural: 3.61 mm
AChE section 168

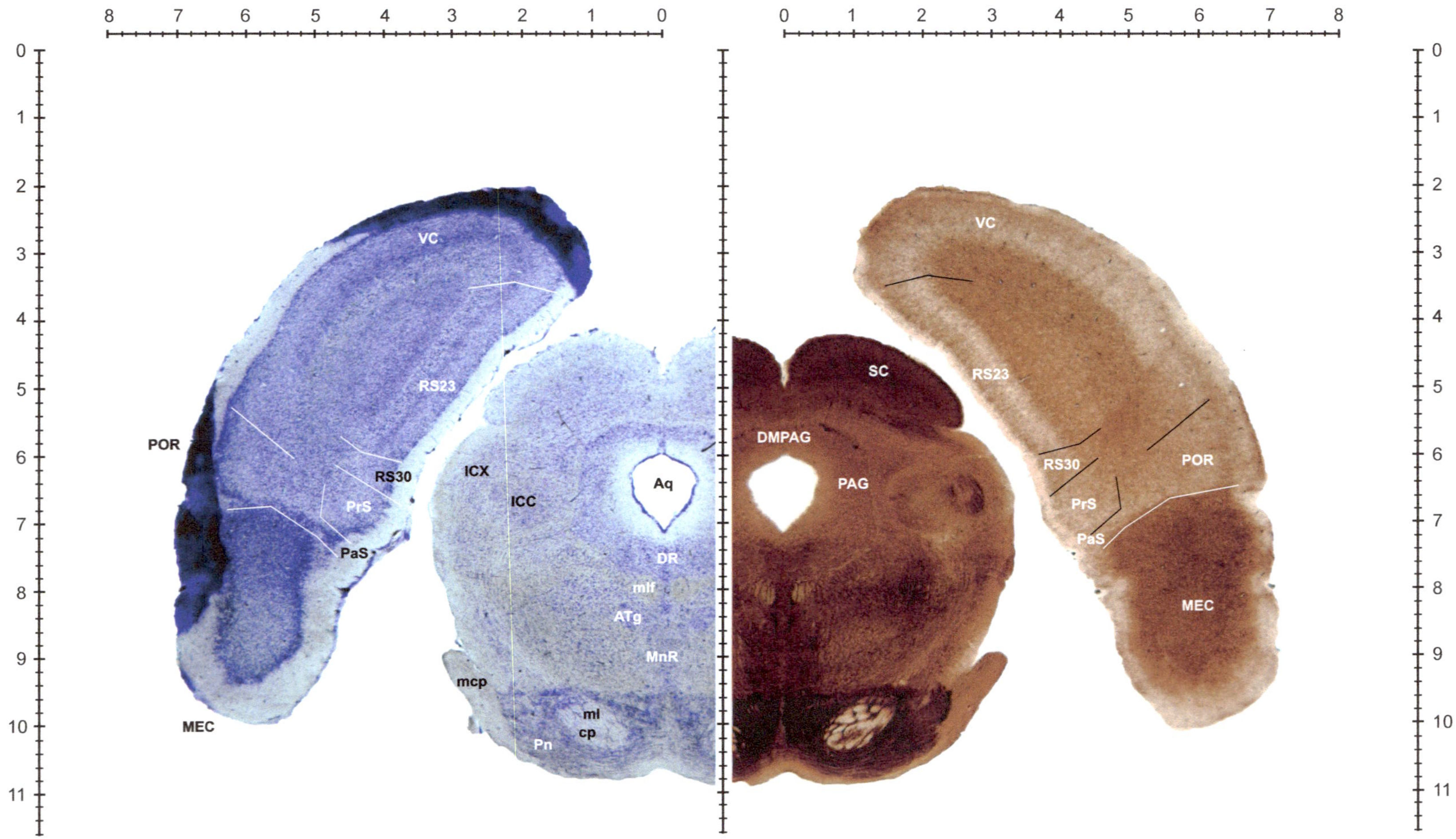


**Aq** aqueduct
**ATg** anterior tegmental nucleus
**cp** cerebral peduncle
**DMPAG** dorsomedial periaqueductal gray
**DR** dorsal raphe nucleus
**ICC** inferior colliculus, central nucleus
**ICX** inferior colliculus, external nucleus
**mcp** middle cerebellar peduncle
**MEC** medial entorhinal cortex
**ml** medial lemniscus
**mlf** medial longitudinal fasciculus
**MnR** median raphe nucleus
**PAG** periaqueductal gray
**PaS** parasubiculum
**Pn** pontine nuclei
**POR** postrhinal cortex
**PrS** presubiculum
**RS23** retrosplenial ctx area 23
**RS30** retrosplenial ctx area 30
**SC** superior colliculus
**VC** visual cortex

## Plate 86 - Sinus: 0.87 mm

Interaural: 3.57 mm

Nissl section 169

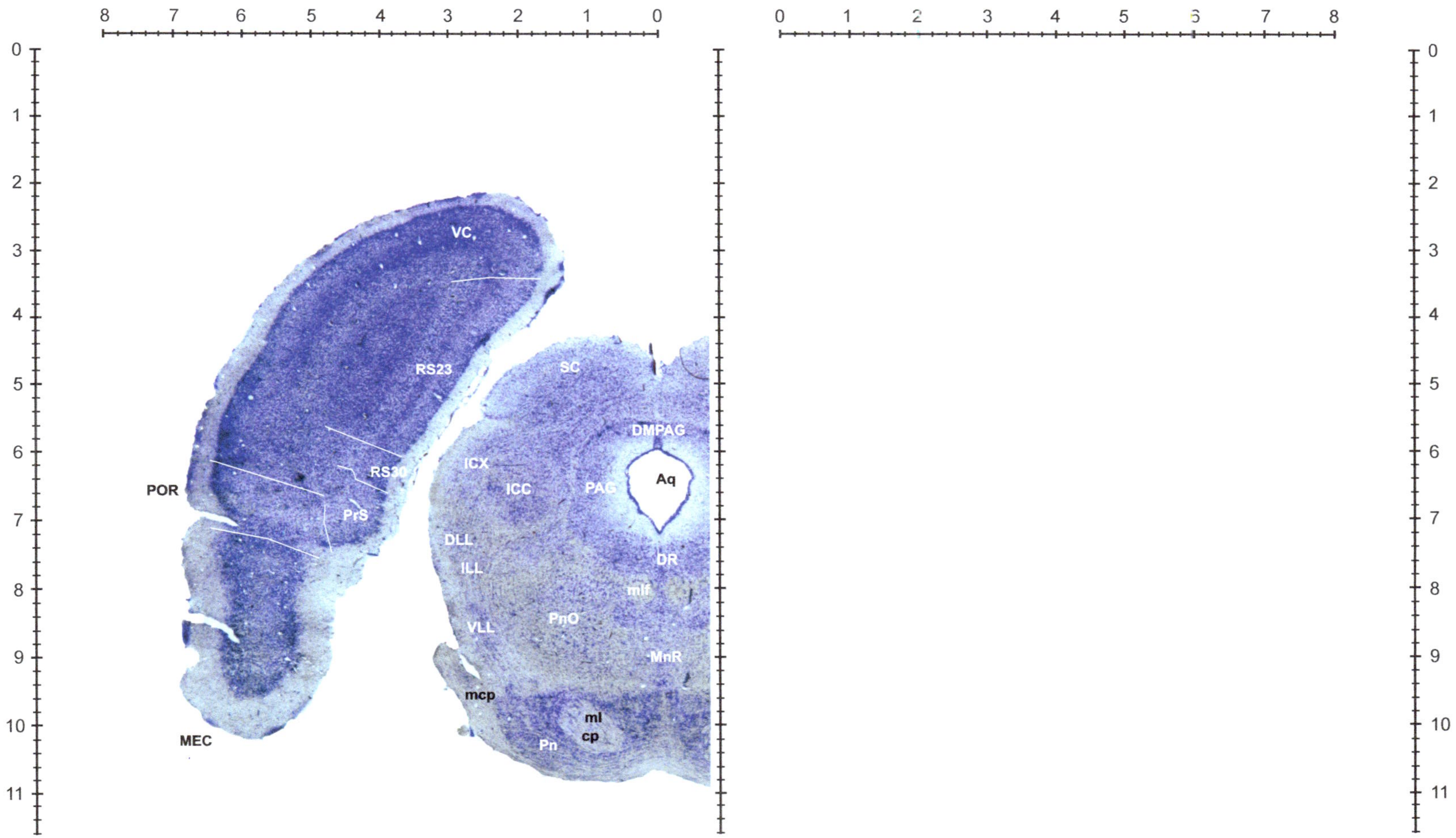


**Aq** aqueduct
**cp** cerebral peduncle
**DLL** dorsal nu of lateral lemniscus
**DMPAG** dorsomedial periaqueductal gray
**DR** dorsal raphe nucleus
**ICC** inferior colliculus, central nucleus
**ICX** inferior colliculus, external nucleus
**ILL** intermediate nu of lateral lemniscus
**mcp** middle cerebellar peduncle
**MEC** medial entorhinal cortex
**ml** medial lemniscus
**mlf** medial longitudinal fasciculus
**MnR** median raphe nucleus
**PAG** periaqueductal gray
**Pn** pontine nuclei
**PnO** pontine reticular nucleus, oral part
**POR** postrhinal cortex
**PrS** presubiculum
**RS23** retrosplenial ctx area 23
**RS30** retrosplenial ctx area 30
**SC** superior colliculus
**VC** visual cortex
**VLL** ventral nu of lateral lemniscus

**Plate 87 - Sinus: 0.71 mm**

Interaural: 3.41 mm

Nissl section 171

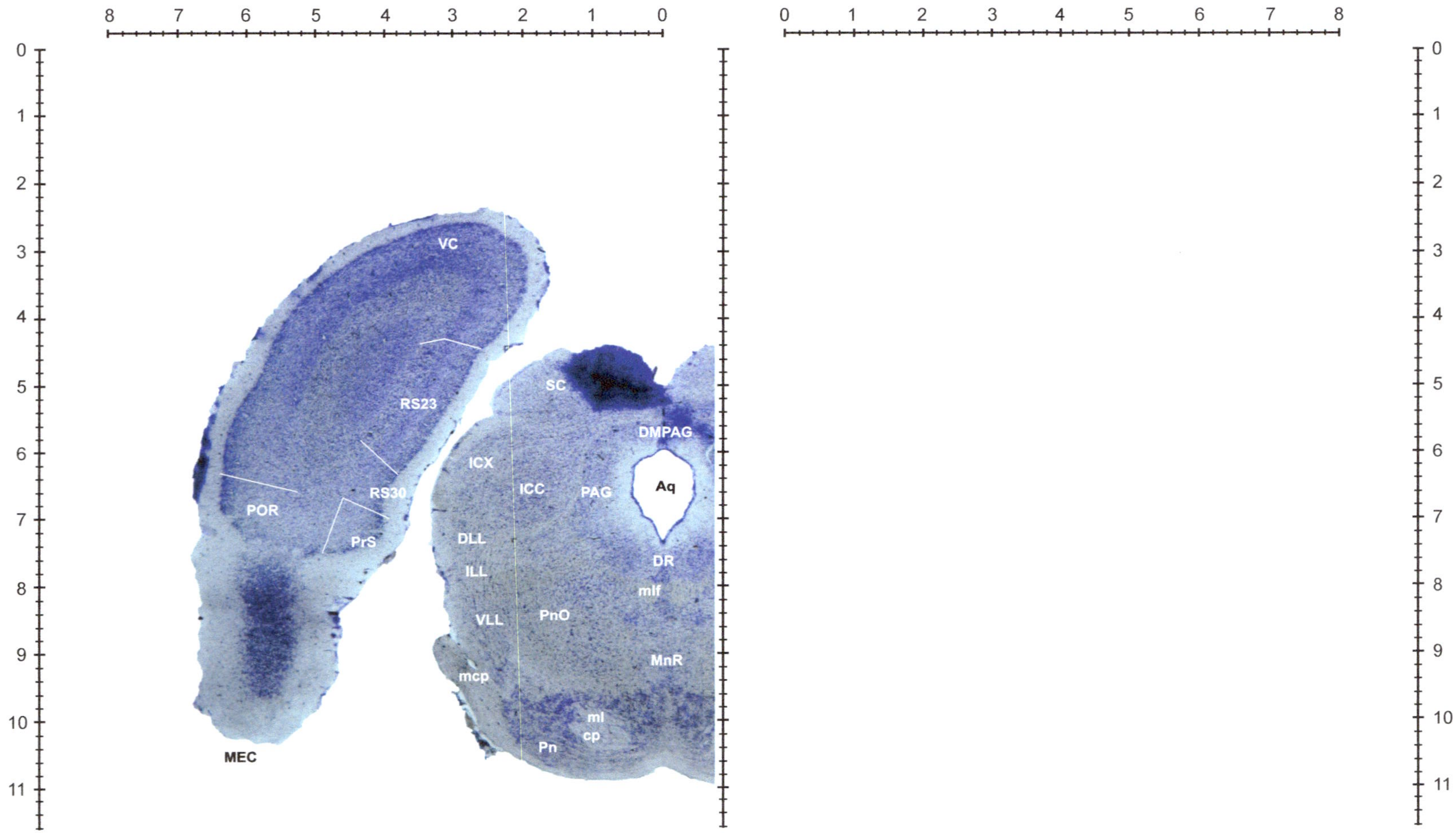


**Aq** aqueduct
**cp** cerebral peduncle
**DLL** dorsal nu of lateral lemniscus
**DMPAG** dorsomedial periaqueductal gray
**DR** dorsal raphe nucleus
**ICC** inferior colliculus, central nucleus
**ICX** inferior colliculus, external nucleus
**ILL** intermediate nu of lateral lemniscus
**mcp** middle cerebellar peduncle
**MEC** medial entorhinal cortex
**ml** medial lemniscus
**mlf** medial longitudinal fasciculus
**MnR** median raphe nucleus
**PAG** periaqueductal gray
**Pn** pontine nuclei
**PnO** pontine reticular nucleus, oral part
**POR** postrhinal cortex
**PrS** presubiculum
**RS23** retrosplenial ctx area 23
**RS30** retrosplenial ctx area 30
**SC** superior colliculus
**VC** visual cortex
**VLL** ventral nu of lateral lemniscus

# Appendix 1

Illustrative sections through the olfactory bulb, cerebellum, and brainstem—taken from the atlas brain. These sections are semistereotaxic: Their scale and medio-lateral coordinates are entirely stereotaxic, as these sections were taken from the atlas brain, and were processed together with the main atlas sections, and we verified that these sections do not contain shrinkage. However, because the olfactory bulb and cerebellum were detached during brain cutting (Methods), we do not have for these sections the correct stereotaxic antero-posterior coordinates and the correct dorso-ventral shift.

## Appendix 1: Semistereotaxic section through the olfactory bulb

Nissl section 8

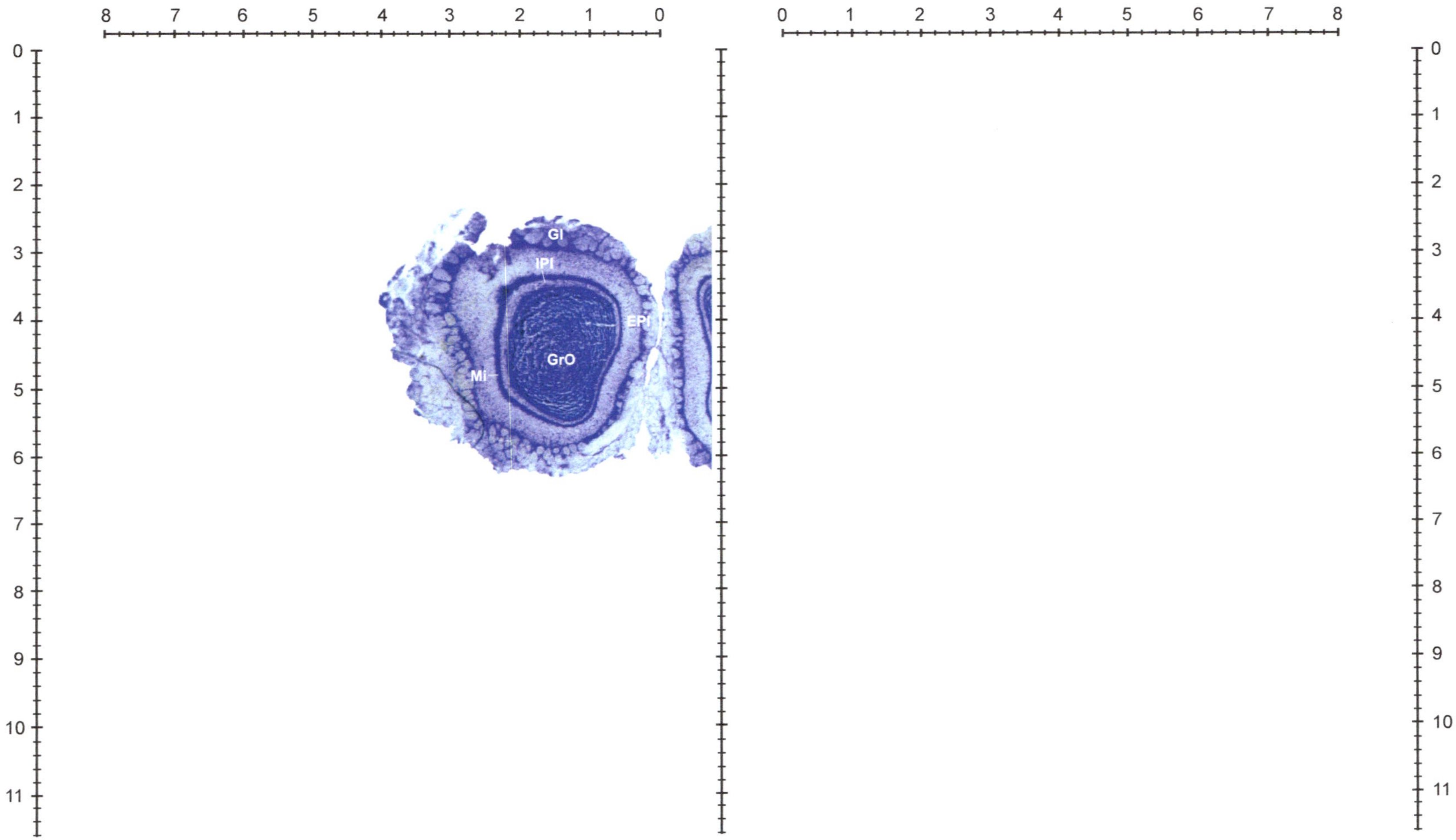


**EPl** external plexiform layer of the olf bulb
**Gl** glomerular layer of the olfactory bulb
**GrO** granule cell layer of the olfactory bulb
**IPl** internal plexiform layer of the olf bulb
**Mi** mitral cell layer of the olfactory bulb

## Appendix 1: Semistereotaxic section through the olfactory bulb

Nissl section 10

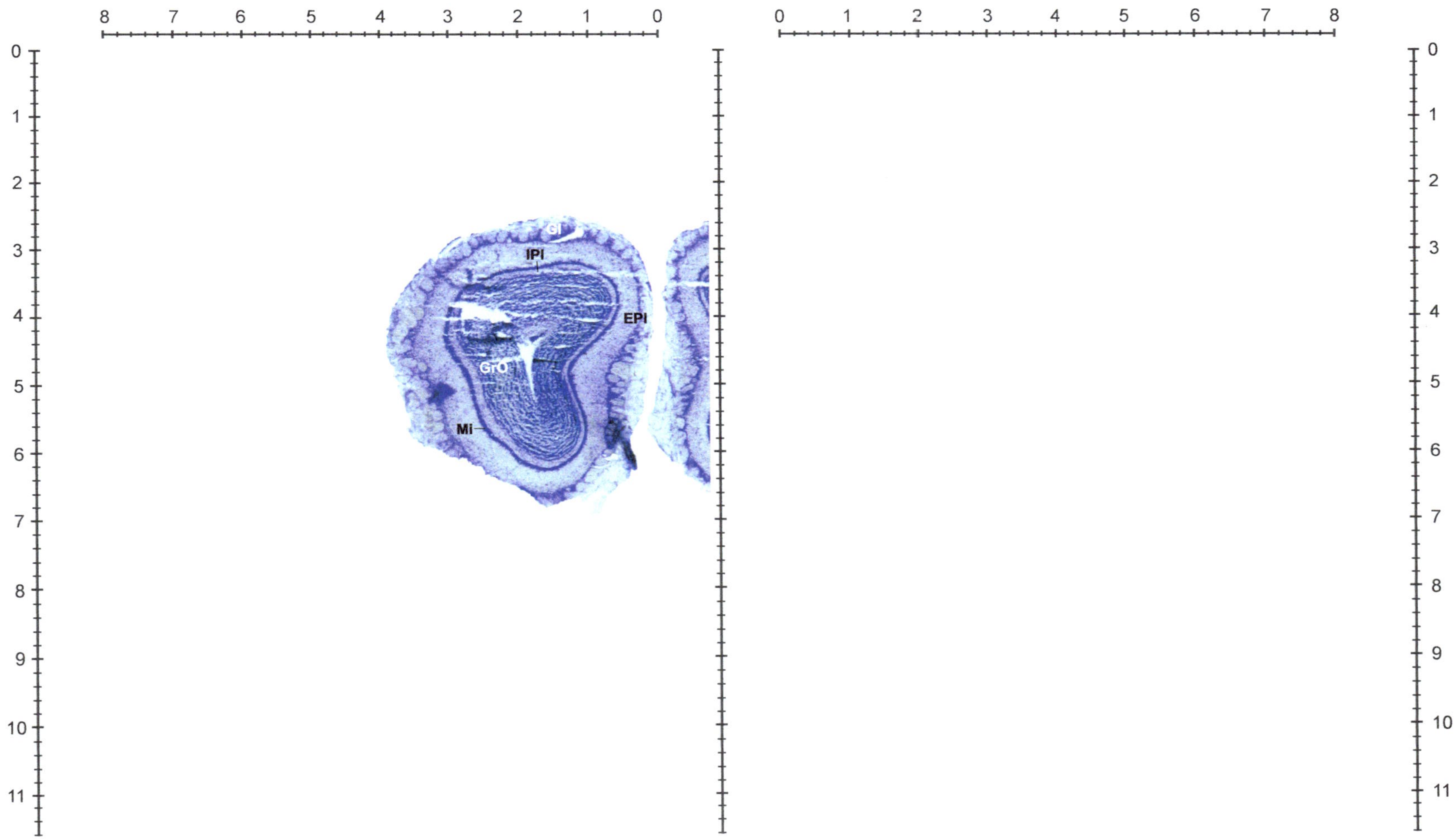


**EPl** external plexiform layer of the olf bulb
**Gl** glomerular layer of the olfactory bulb
**GrO** granule cell layer of the olfactory bulb
**IPl** internal plexiform layer of the olf bulb
**Mi** mitral cell layer of the olfactory bulb

## Appendix 1: Semistereotaxic section through the cerebellum and brainstem

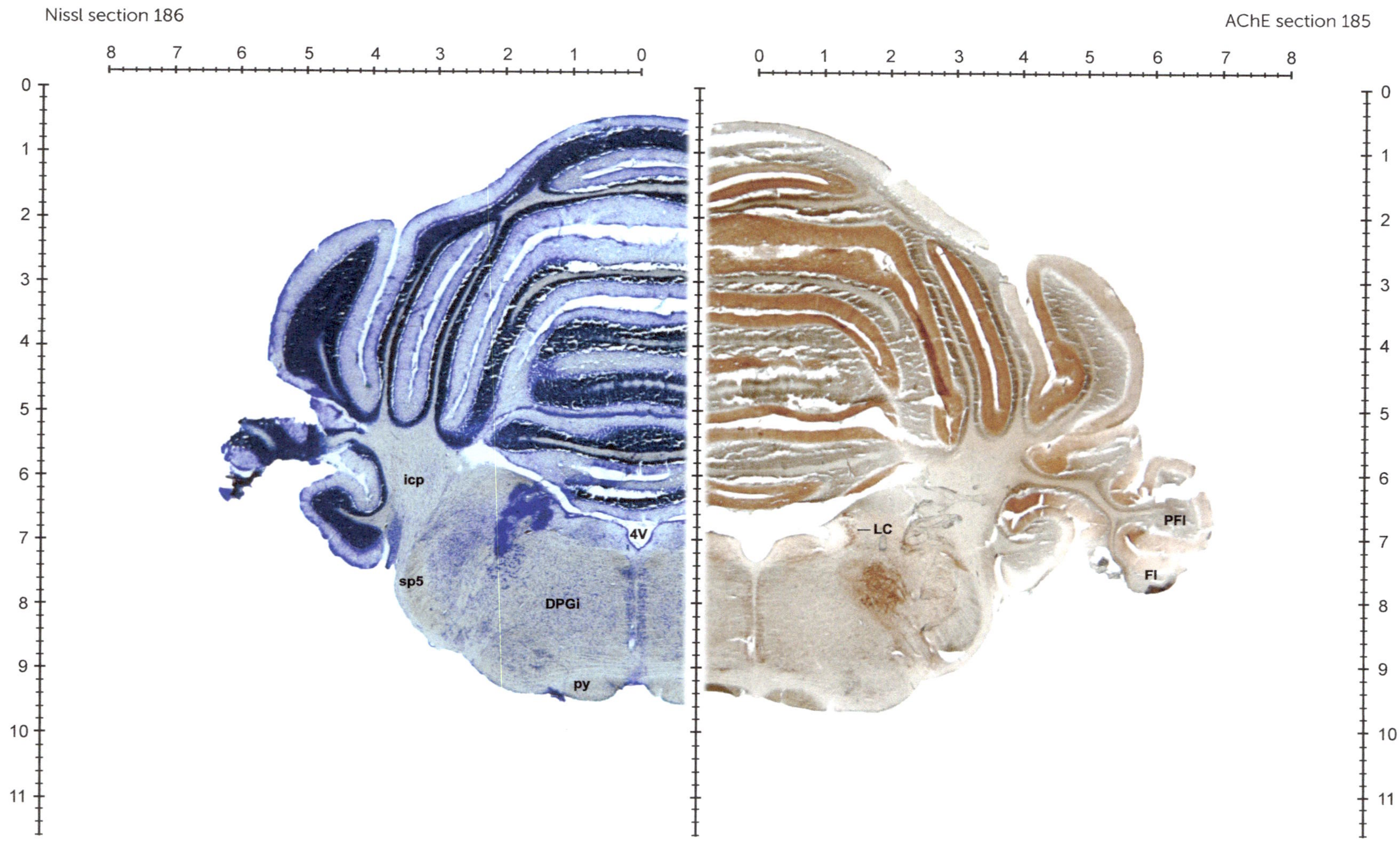


**4V** 4th ventricle
**DPGi** dorsal paragigantocellular nucleus
**Fl** flocculus
**icp** inferior cerebellar peduncle
**LC** locus coeruleus
**PFl** paraflocculus
**py** pyramidal tract
**sp5** spinal trigeminal tract

## Appendix 1: Semistereotaxic section through the cerebellum and brainstem

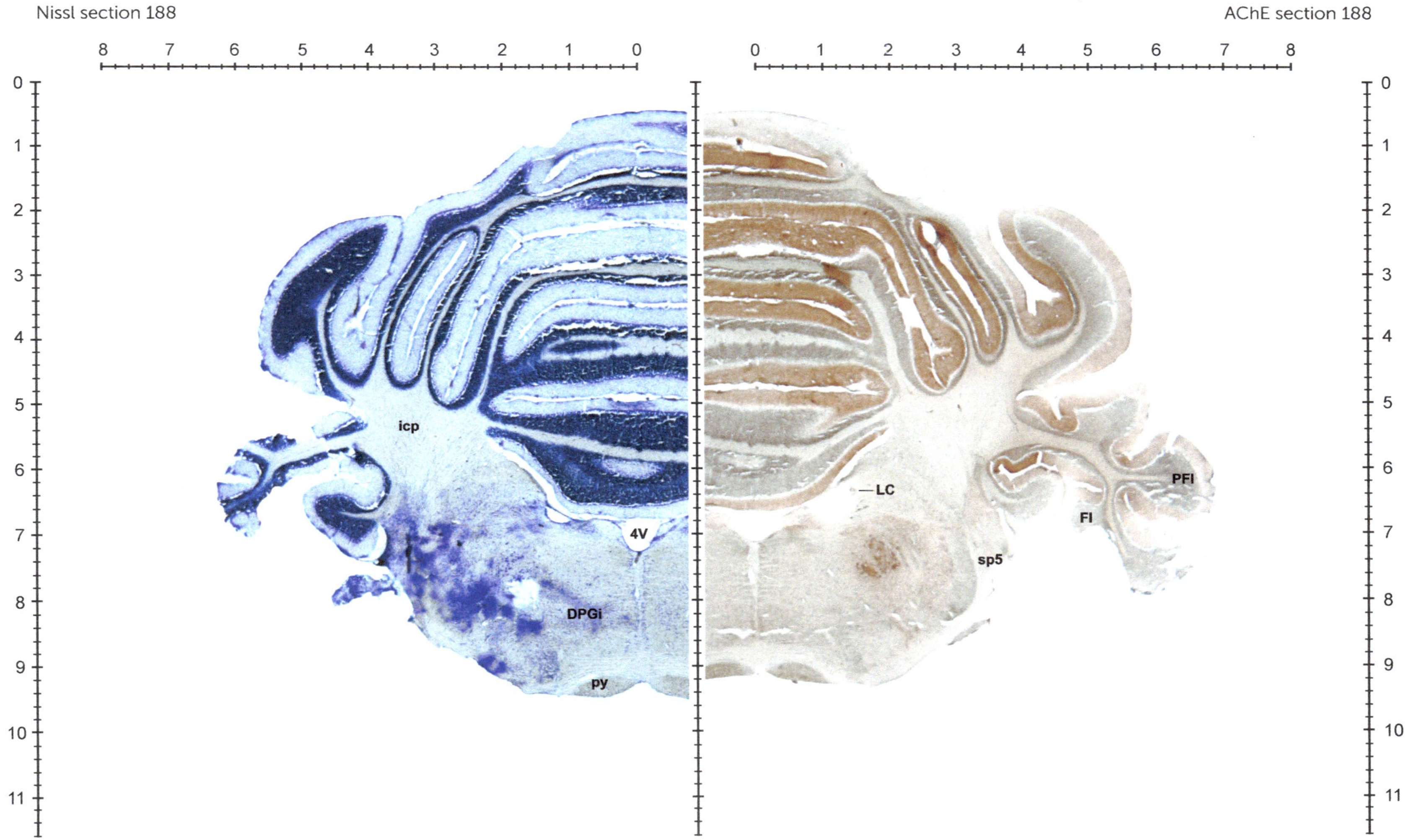


**4V** 4th ventricle
**DPGi** dorsal paragigantocellular nucleus
**Fl** flocculus
**icp** inferior cerebellar peduncle
**LC** locus coeruleus
**PFl** paraflocculus
**py** pyramidal tract
**sp5** spinal trigeminal tract

# Appendix 2

Tyrosine hydroxylase (TH) staining of coronal sections taken not from the atlas brain but from another individual bat (male). Shown are sections at approximately 0.61 mm intervals. The first page shows all the 16 sections at low magnification, and the subsequent 4 pages show 4 sections each at a higher magnification. The approximate plate number in the atlas is marked on each image (e.g., P-17, P-83). Note that the cutting was not fully coronal here, but was slightly tilted; we determined the approximate atlas plate based on the similarity of the ventral half of these sections to the atlas plates—we focused on the ventral half as it is the most relevant for TH staining, since it aids in identifying dopamine-rich regions, such as VTA, SNR, SNC, mfb, etc., which are located ventrally. Scale bars were corrected here for shrinkage, to match the in vivo scale of the atlas. We marked several identified brain regions on these sections, using the same notations as in the atlas plates.

## List of structures marked in this appendix:

**C** caudate
**DR** dorsal raphe nucleus
**GP** globus pallidus
**mfb** medial forebrain bundle
**Pu** putamen
**PVH** paraventricular hypothalamic nucleus
**SNC** substantia nigra, compact part
**SNR** substantia nigra, reticular part
**VTA** ventral tegmental area

## Appendix 2: TH staining - coronal sections through the brain

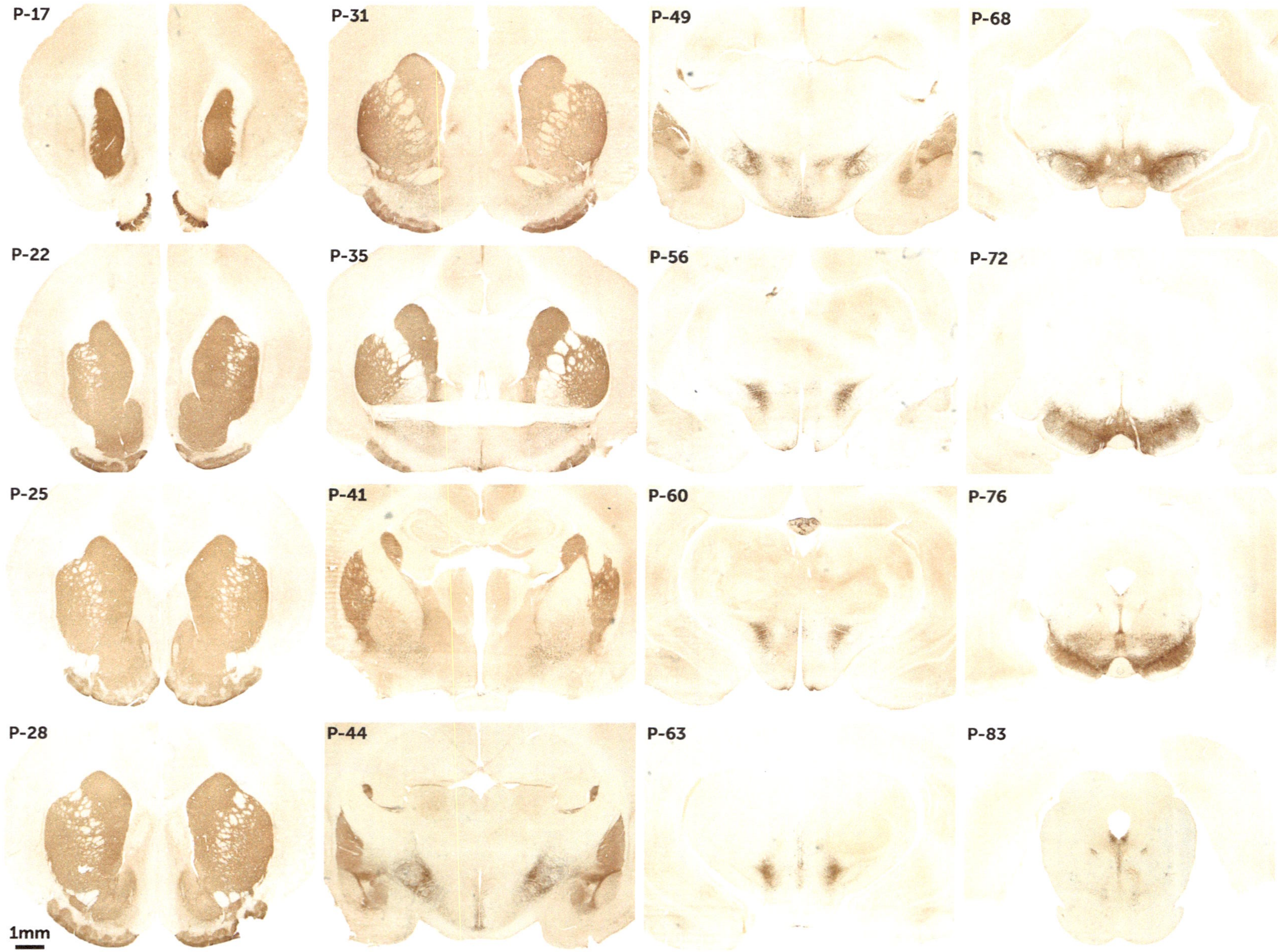

## Appendix 2: TH staining - coronal sections through the brain

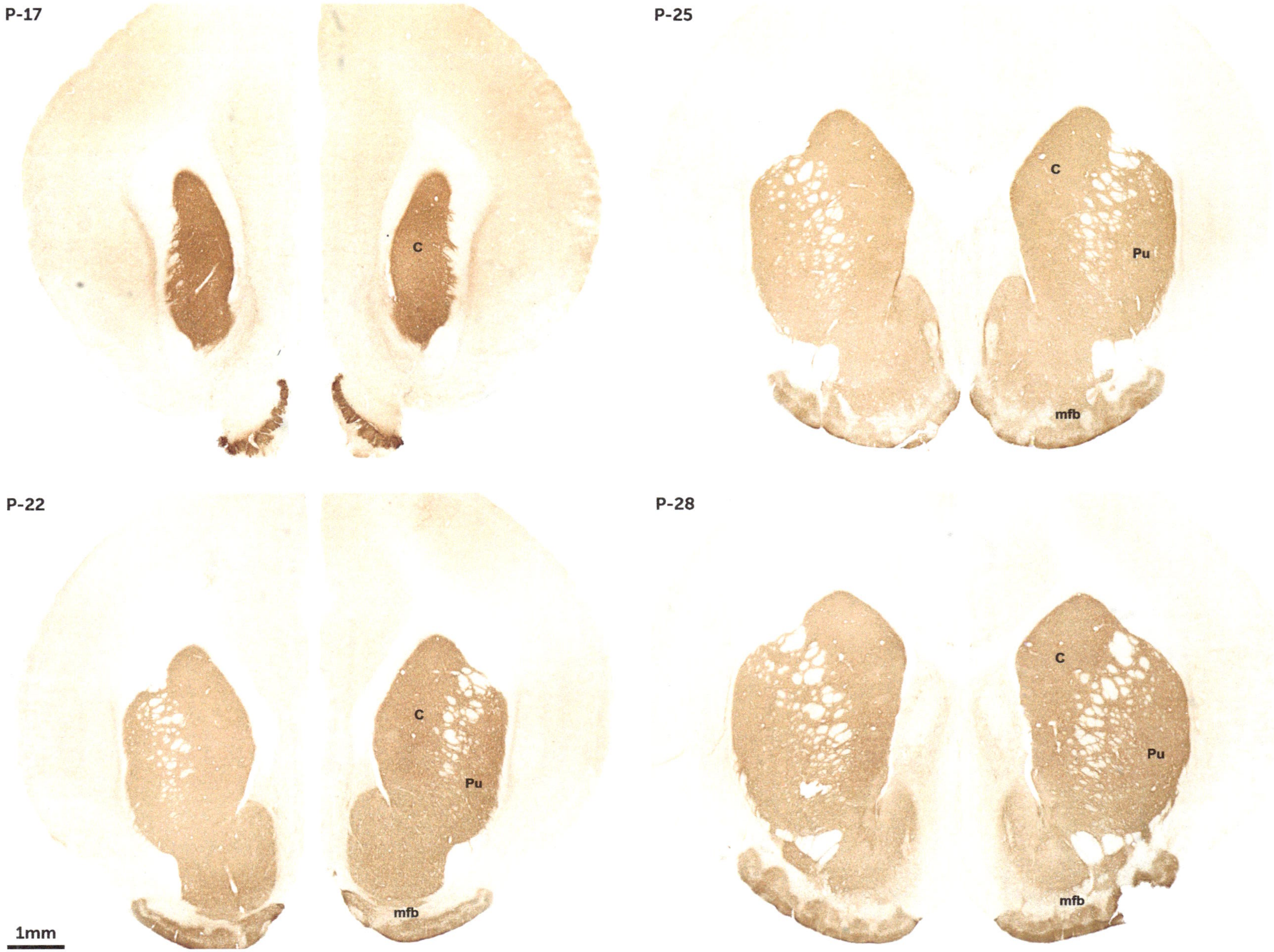

## Appendix 2: TH staining - coronal sections through the brain

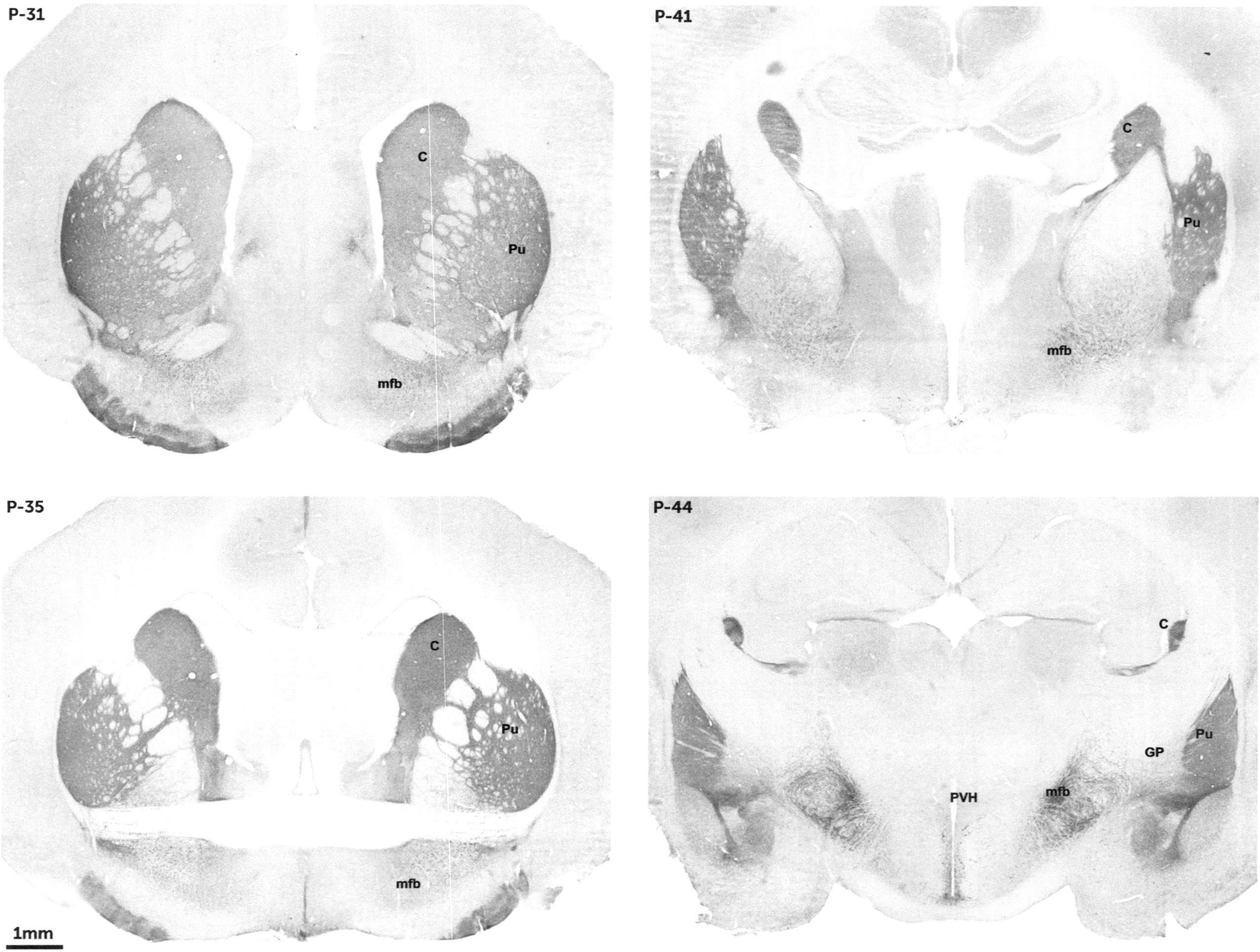

## Appendix 2: TH staining - coronal sections through the brain

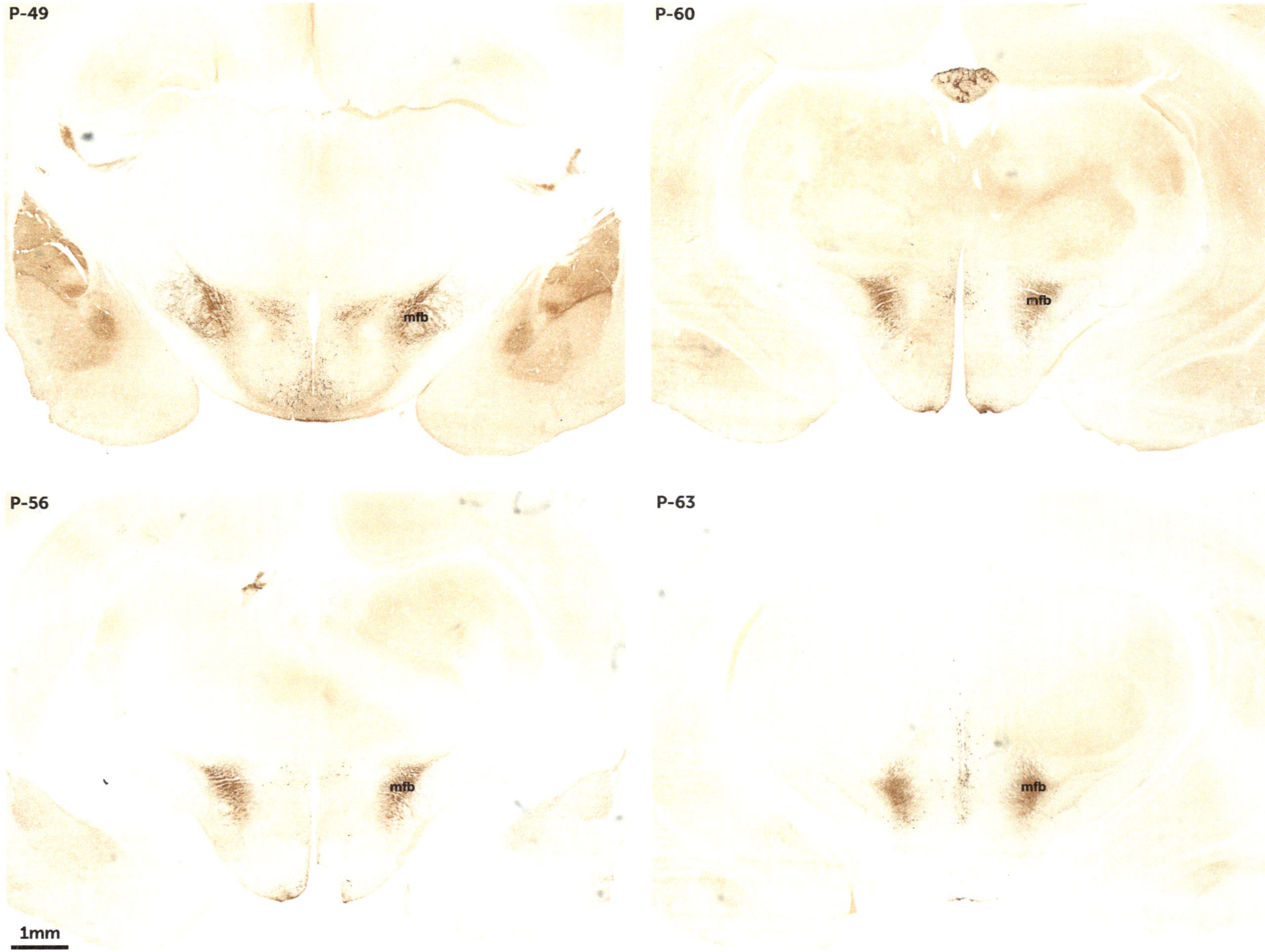

## Appendix 2: TH staining - coronal sections through the brain

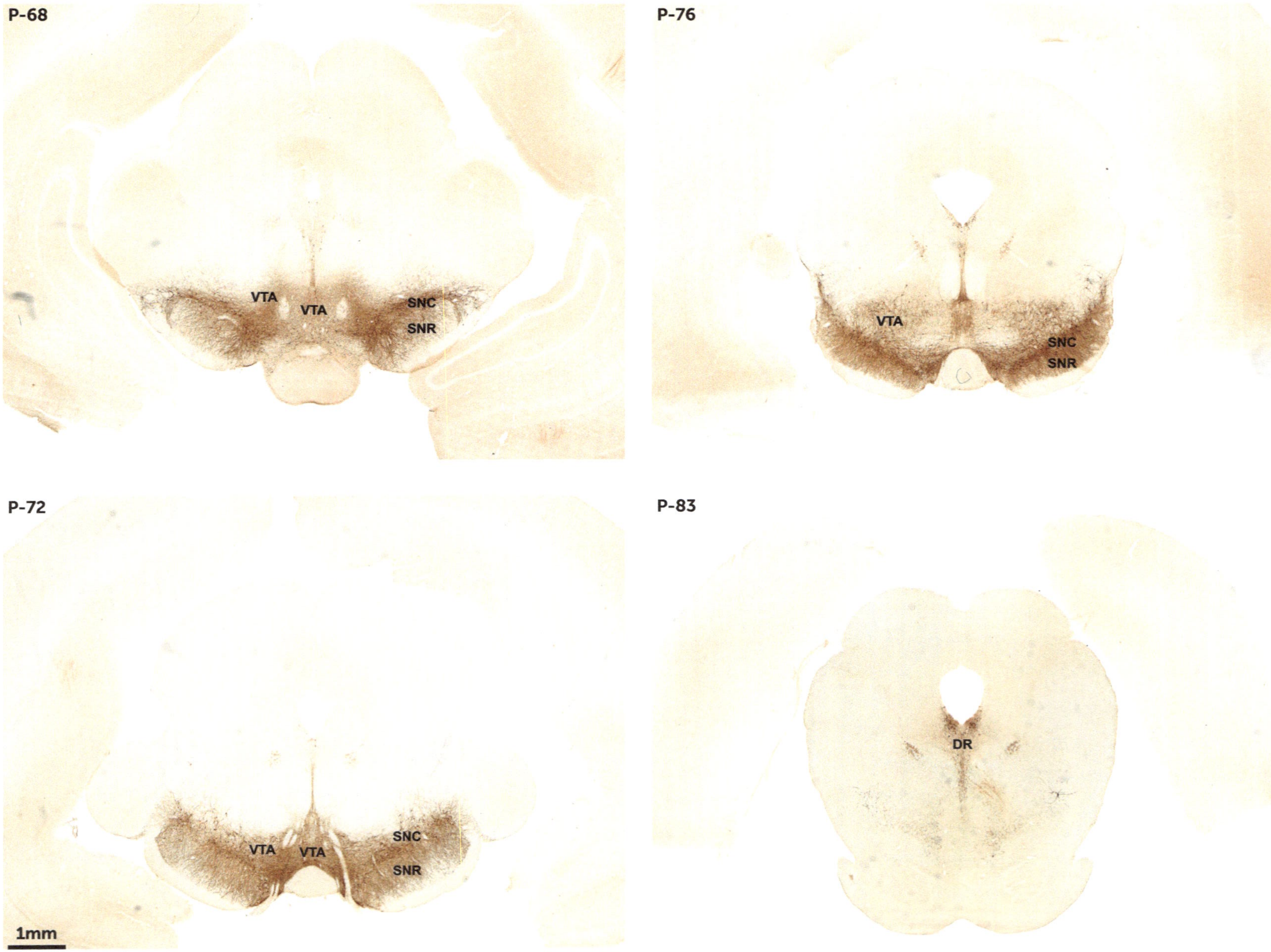

# Appendix 3

Double staining by choline acetyltransferase (ChAT, red) and tyrosine hydroxylase (TH, green) of coronal sections, taken from another individual bat (male)—focusing on the major cholinergic nuclei of the brain (see Methods, "Basal ganglia and basal forebrain"). Top panels show low-magnification images; bottom panels show high-magnification images of the regions marked by white rectangles at the top. The approximate plate number in the atlas is marked on each image (e.g., P-31, P-44). Note that the cutting was not fully coronal here, but was slightly tilted; we determined the approximate atlas plate based on the similarity of the ventral half of these sections to the atlas plates—we focused on the ventral half as it is the most relevant for ChAT staining, and it aids in identifying cholinergic-rich regions, such as GP, VP, nucleus basalis, and HDB, which are located ventrally. Scale bars were corrected here for shrinkage, to match the in vivo scale of the atlas. We marked several identified brain regions on these sections, using the same notation as in the atlas plates.

## List of structures marked in this appendix:

**B** basal nucleus (nucleus basalis)
**GP** globus pallidus
**HDB** horizontal limb nu diagonal band
**mfb** medial forebrain bundle
**Pu** putamen
**VP** ventral pallidum

## Appendix 3: ChAT + TH fluorescent images of coronal sections through the basal forebrain

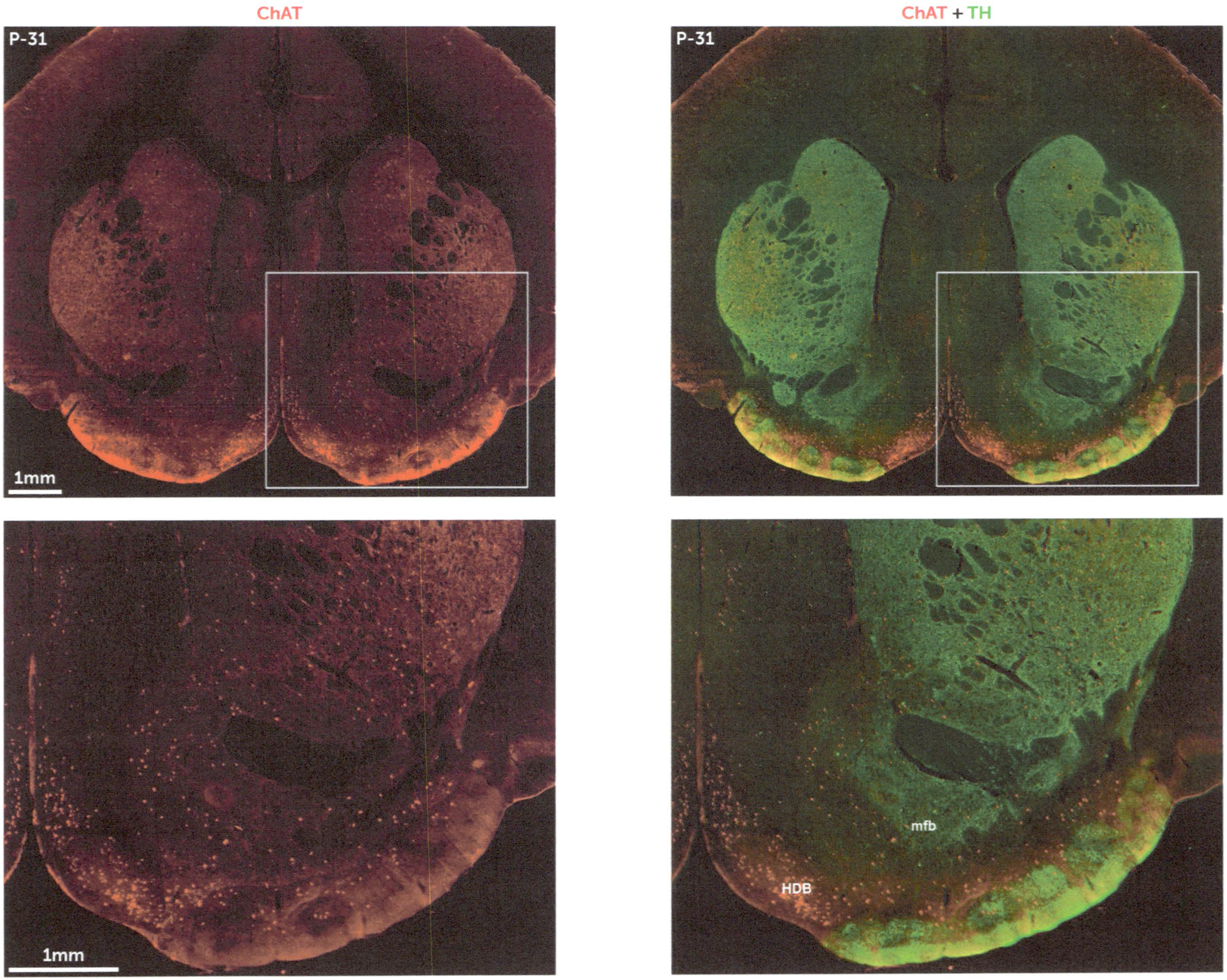

## Appendix 3: ChAT + TH fluorescent images of coronal sections through the basal forebrain

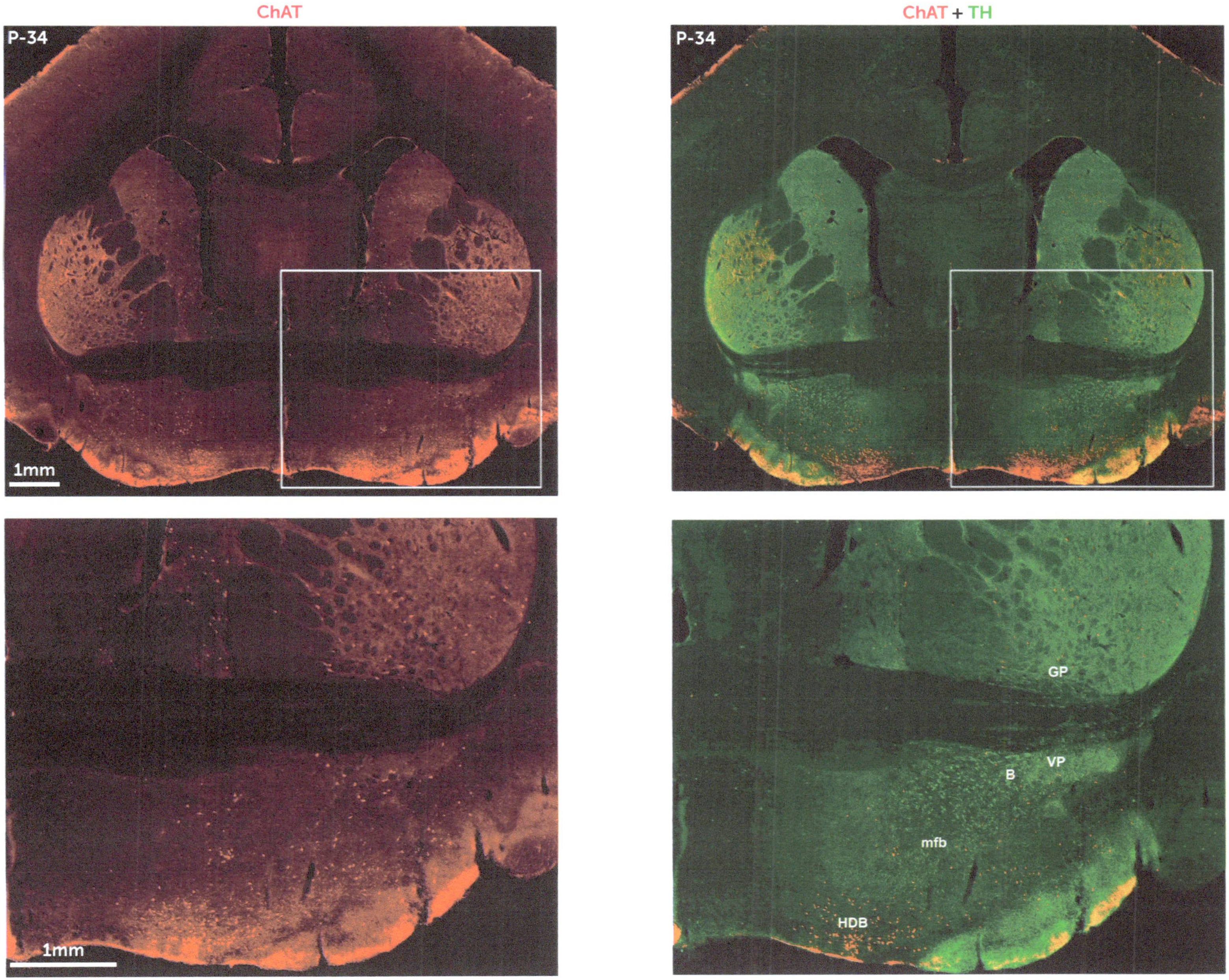

## Appendix 3: ChAT + TH fluorescent images of coronal sections through the basal forebrain

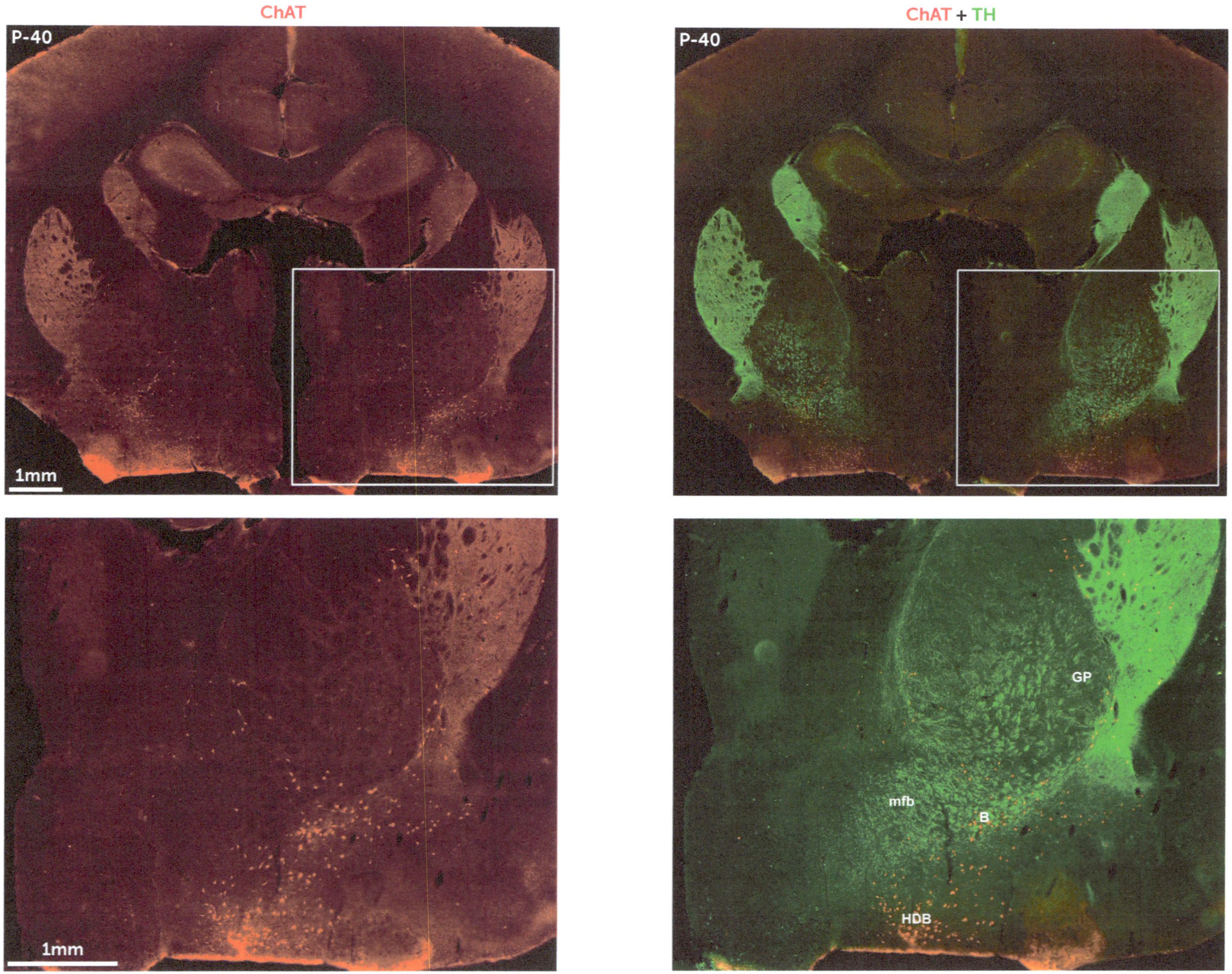

## Appendix 3: ChAT + TH fluorescent images of coronal sections through the basal forebrain

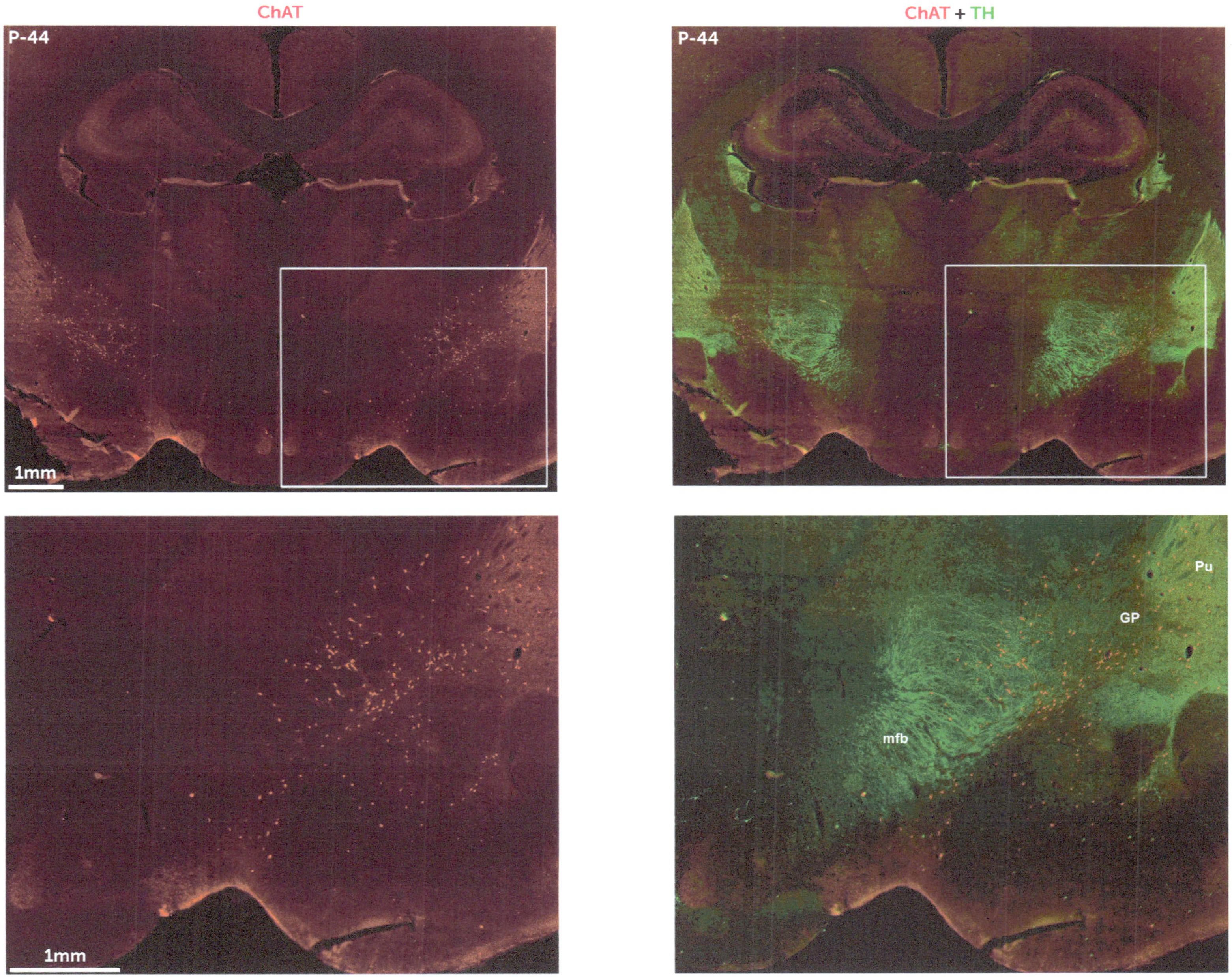

# Appendix 4

Oxytocin staining of a coronal section through the hypothalamus, taken not from the atlas brain but from another individual bat (male) (see Methods, "Oxytocin immunohistochemistry"). The approximate plate number in the atlas is marked (P-44). Scale bars were corrected here for shrinkage, to match the in vivo scale of the atlas. We marked several identified brain regions on these sections, using the same notation as in the atlas plates.

## List of structures marked in this appendix:

**3V** 3rd ventricle
**PVH** paraventricular hypothalamic nucleus
**SO** supraoptic nucleus

## Appendix 4: Oxytocin - Fluorescent image of coronal section through the Hypothalamus

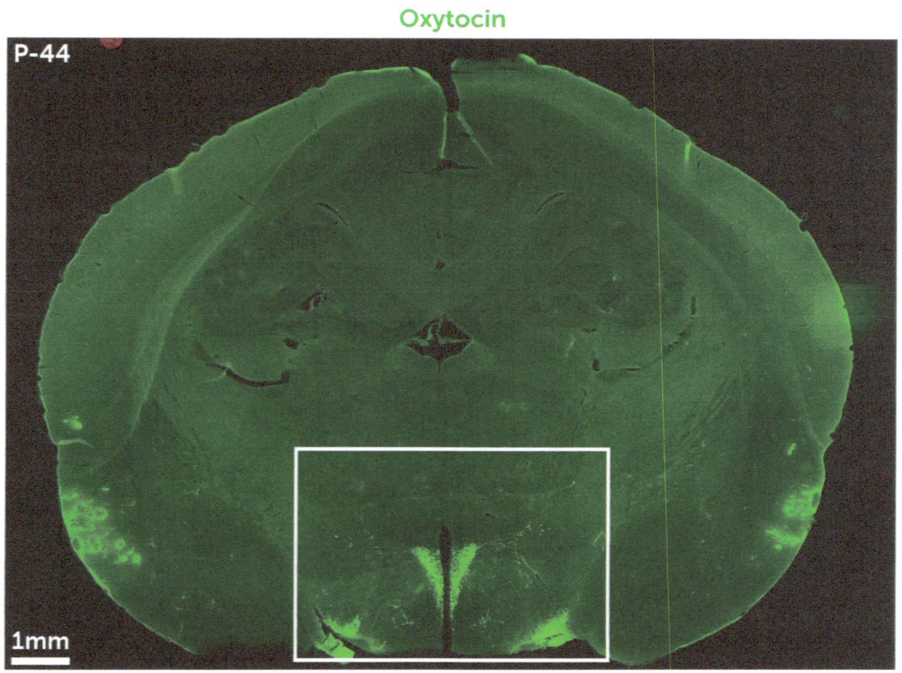


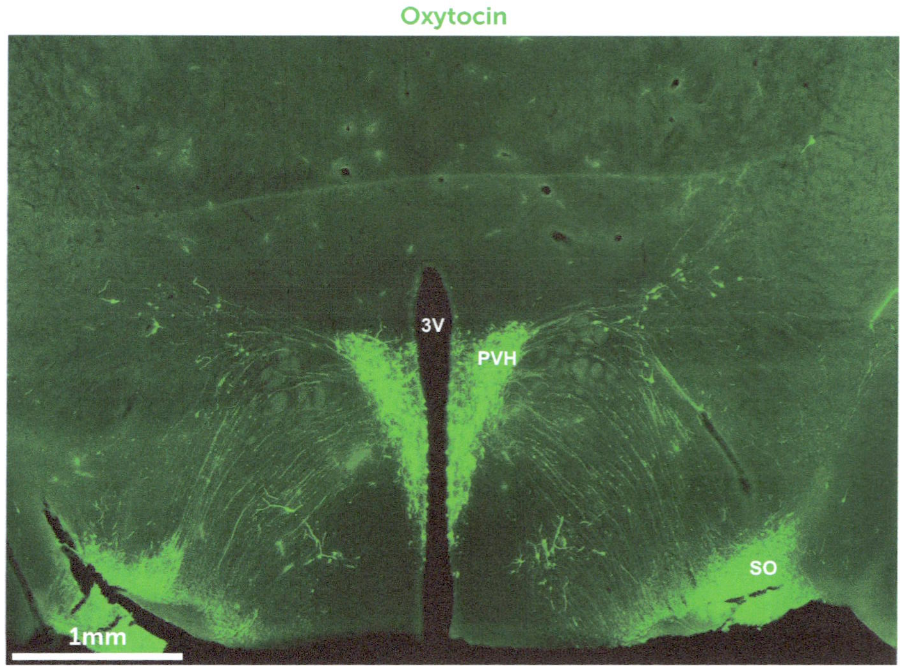

# Appendix 5

Tracer injections that were used to delineate the auditory cortex and auditory thalamus (see details in Methods, "Injections of tracers into the auditory thalamus and auditory cortex"). The first page shows injection of biotinylated dextran amine (BDA, anterograde tracer) into auditory thalamus nucleus MGV (top-left), revealing staining in primary auditory cortex, Au1 (bottom-left); and injection of BDA into auditory thalamus nuclei MGM and MGD of another bat (top-right), revealing staining in the secondary auditory cortical ventral field, AuV (bottom-right). The second page shows double injection of BDA and cholera toxin (CTx, retrograde tracer), with both injections placed into AuV (top)—revealing connectivity with auditory thalamus nuclei MGM and MGD (bottom); the rightmost images show high-magnification images of the regions marked by white rectangles on the left. The tracers were photographed either using fluorescent light or with a combination of fluorescent light and darkfield illumination. Cholera toxin tracer was photographed through a Rhodamin filter, and BDA tracer was photographed through an FITC filter. These images (and additional images from several other bats injected with BDA and cholera toxin into auditory thalamus and auditory cortex) were used to identify Au1 and AuV and mark them in the atlas.

## List of structures mentioned in this appendix:

**Au1** primary auditory cortex
**AuV** secondary auditory ctx ventral area
**MGD** med geniculate nucleus, dorsal part
**MGM** med geniculate nucleus, medial part
**MGV** med geniculate nucleus, ventral part

**Appendix 5: Auditory system - Injection of BDA into thalamus MGV revealing staining in Au1 (left) and injection into thalamus MGM and MGD revealing staining in AuV (right)**

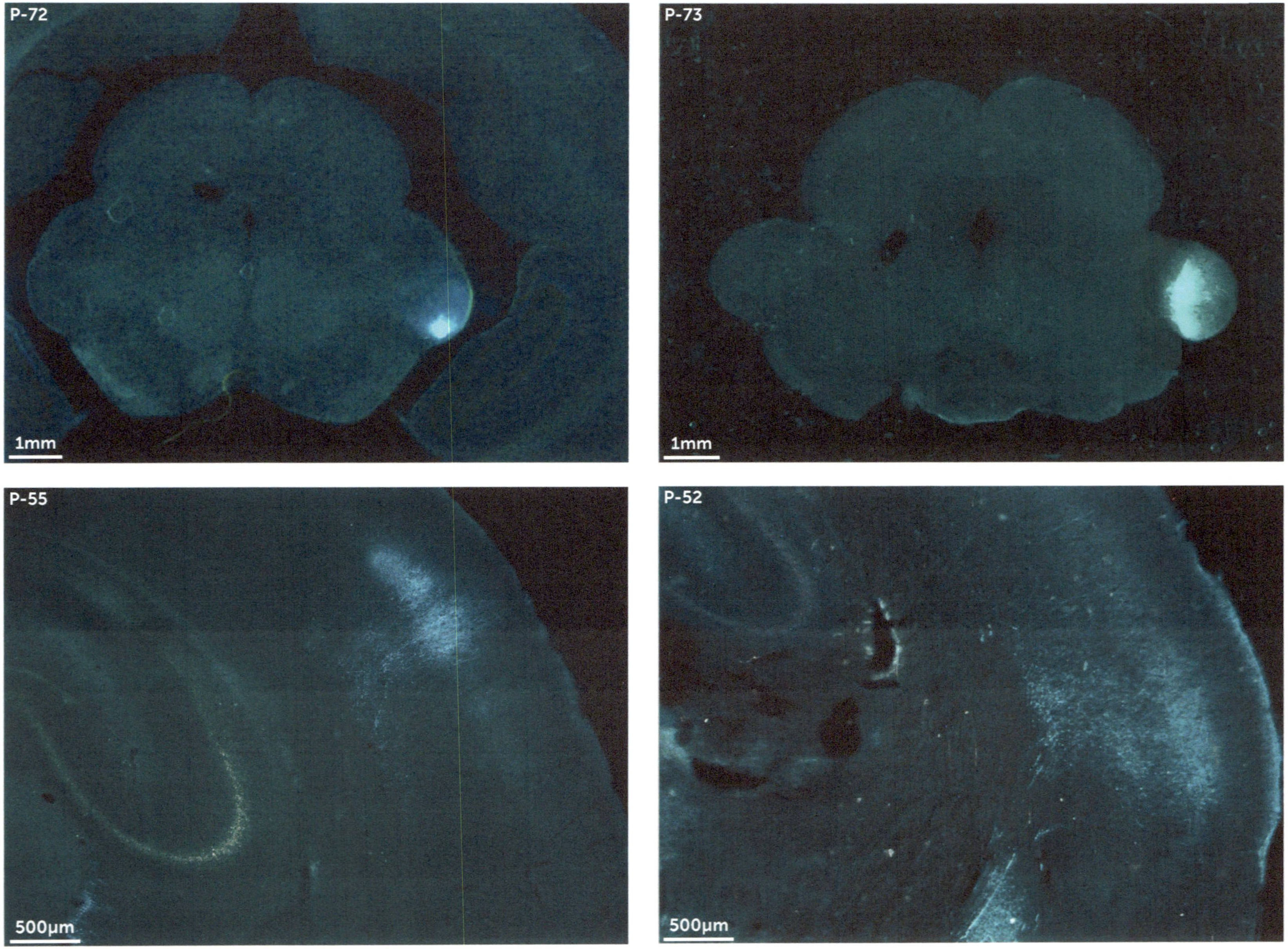

**Appendix 5: Auditory system - double injection of BDA and cholera toxin (CTx) into AuV (top), revealing connectivity with auditory thalamus nuclei MGD and MGM (bottom)**

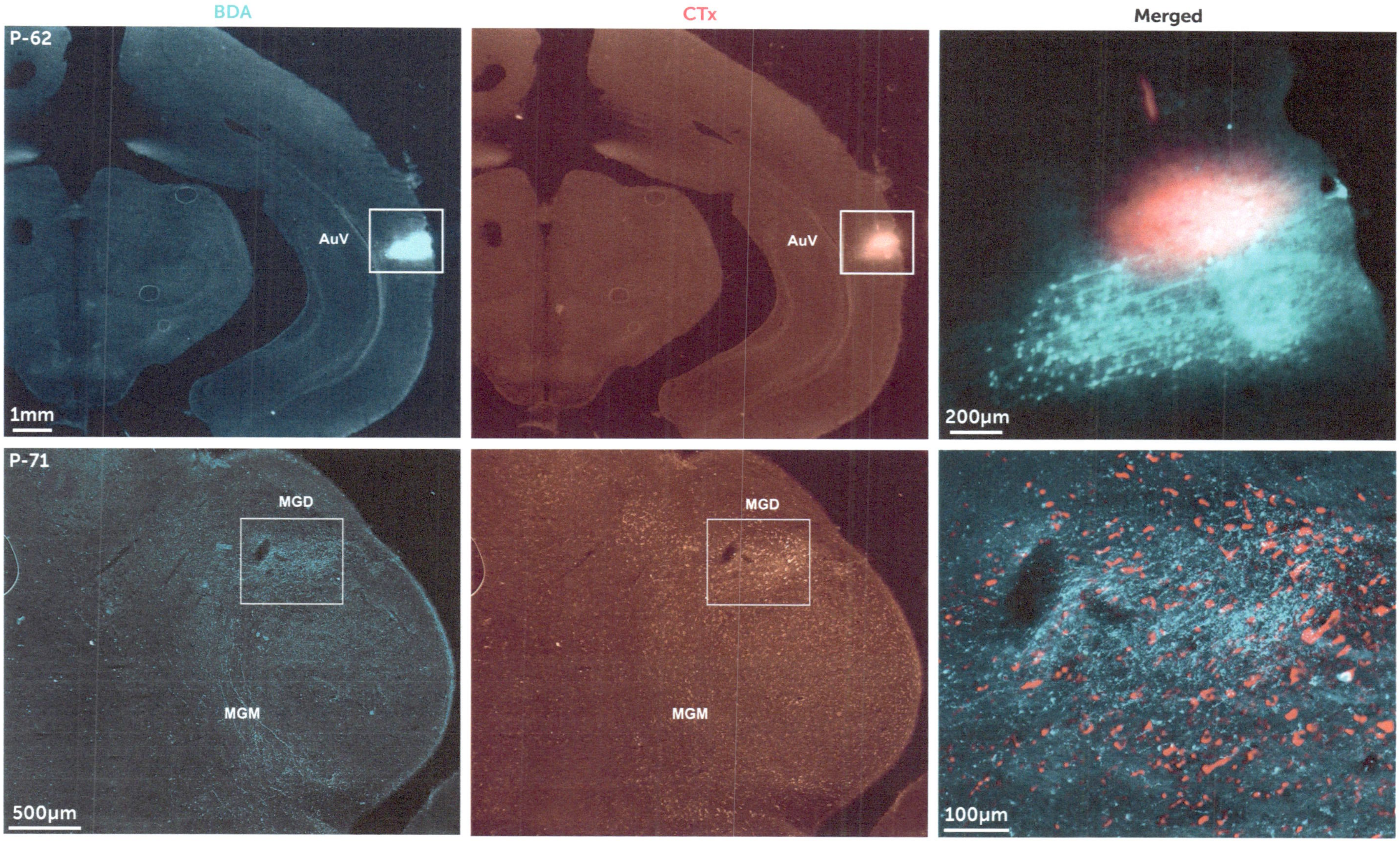

# Appendix 6

3D views. (A–D) 3D reconstruction images of the hippocampal formation and surrounding regions (see Methods, "Hippocampal region"). DG, dentate gyrus; CA1, CA2, CA3, the CA fields; SUB, subiculum; PrS, presubiculum; PaS, parasubiculum; MEC, medial entorhinal cortex; LEC, lateral entorhinal cortex; POR, postrhinal cortex; PER, perirhinal cortex (subdivided in the atlas into PRh35 and PRh36); HF, hippocampal formation (comprising DG, CA fields, and SUB); PHR, parahippocampal region (comprising PrS, PaS, MEC, LEC, POR, and PER). Reconstruction was done in Neurolucida, based on coronal sections (as in the atlas plates), as well as sagittal and horizontal sections (as in Appendixes 7 and 8), where we verified consistency of borders between all three planes of section. The four panels A–D show the same 3D volume from four different angles. The cardinal directions are marked (D, dorsal; V, ventral; M, medial; L, lateral; R, rostral; C, caudal). We note that we have full MRI volumetric data for the brain of the Egyptian fruit bat, and we utilized the MR volume for doing the 3D reconstructions in this Appendix. (E) Dissection of a brain of an Egyptian fruit bat, showing the 3D position of the hippocampus.

## Appendix 6: 3D views of the hippocampal formation and surrounding regions

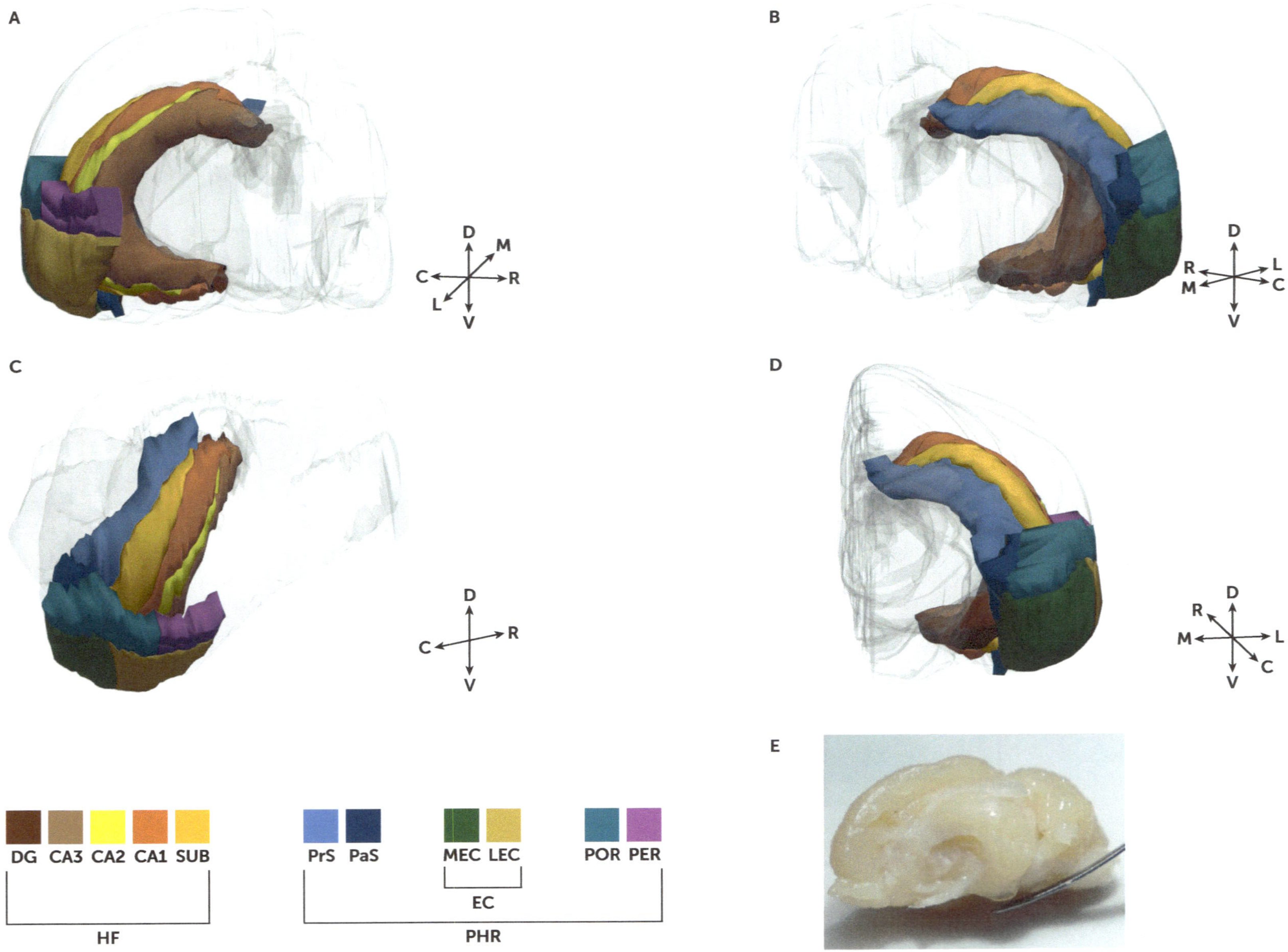

# Appendix 7

Series of sagittal sections through the brain of another bat (female), alternating Nissl and Parvalbumin (PV) staining. The first page shows all the 12 sections at low magnification, and the subsequent 3 pages show 4 sections each at a higher magnification. The approximate M-L (medio-lateral) atlas stereotaxic coordinates are indicated.

## Appendix 7: Sagittal sections throughout the brain – Nissl and PV staining

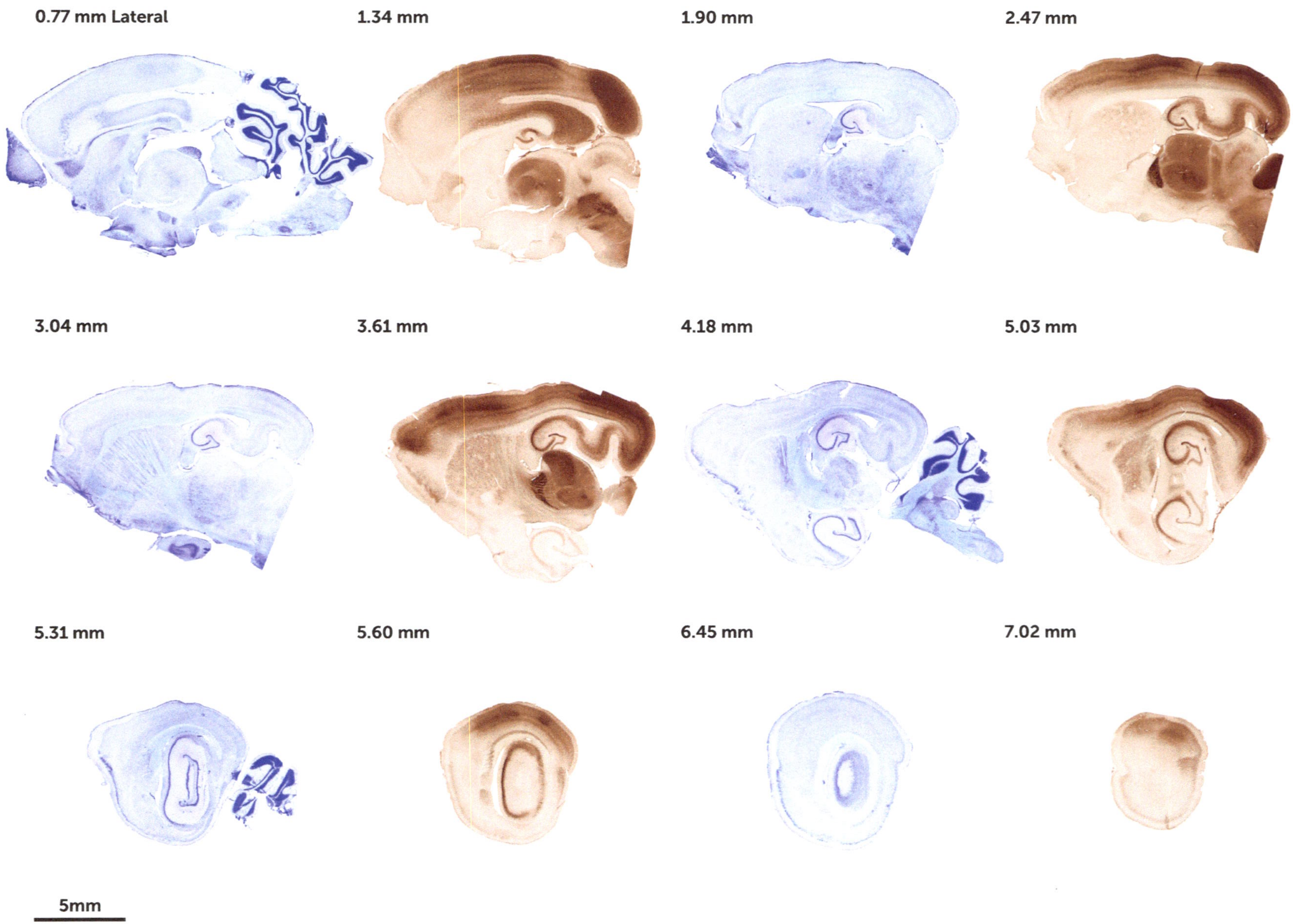

## Appendix 7: Sagittal sections throughout the brain – Nissl and PV staining

0.77 mm Lateral

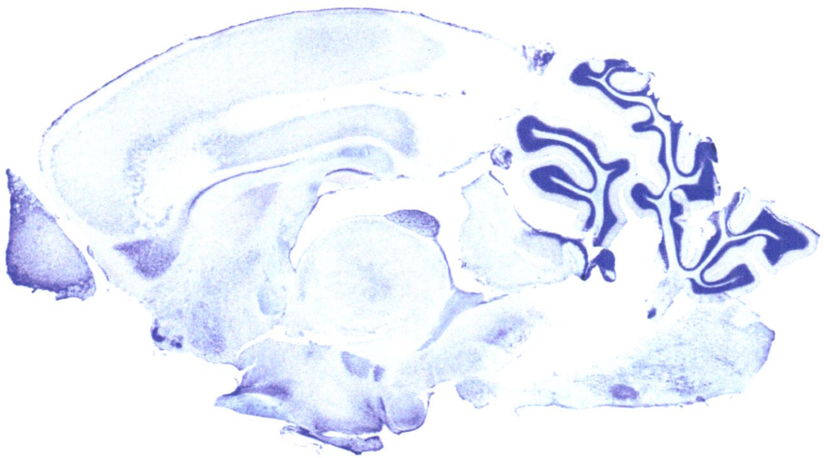

1.34 mm

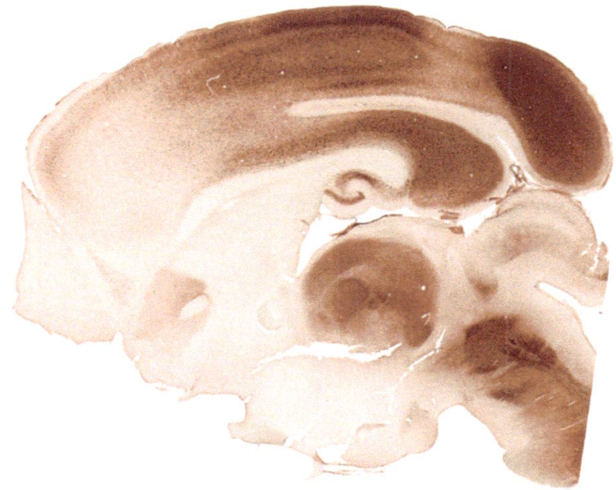

1.90 mm

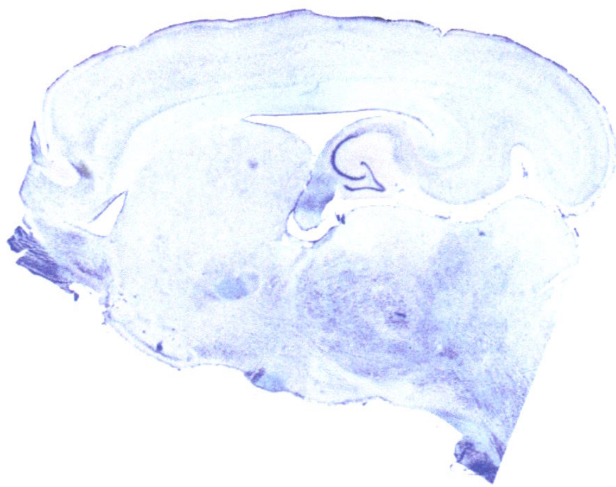

2.47 mm

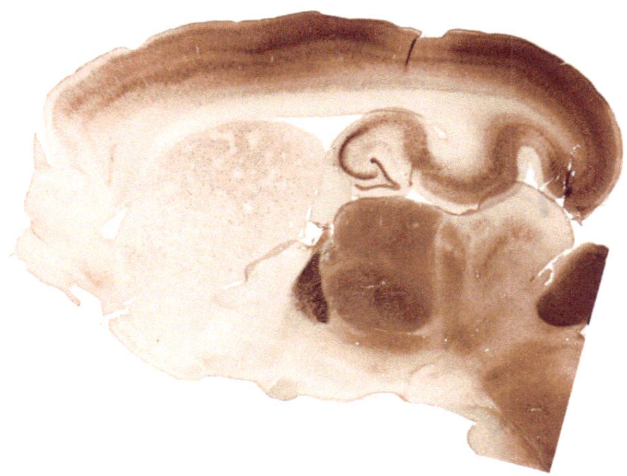

2mm

## Appendix 7: Sagittal sections throughout the brain – Nissl and PV staining

**3.04 mm Lateral**

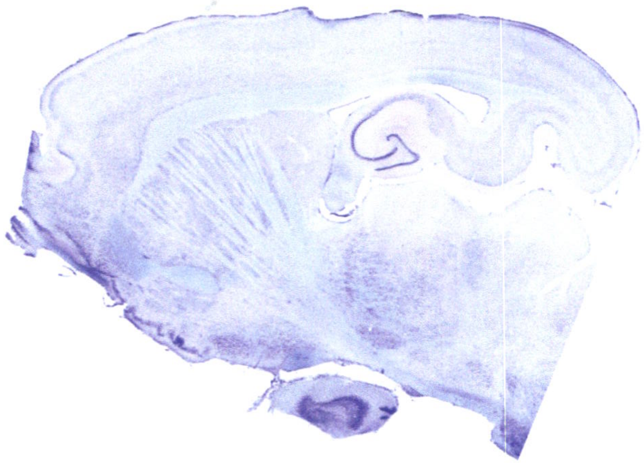

**3.61 mm**

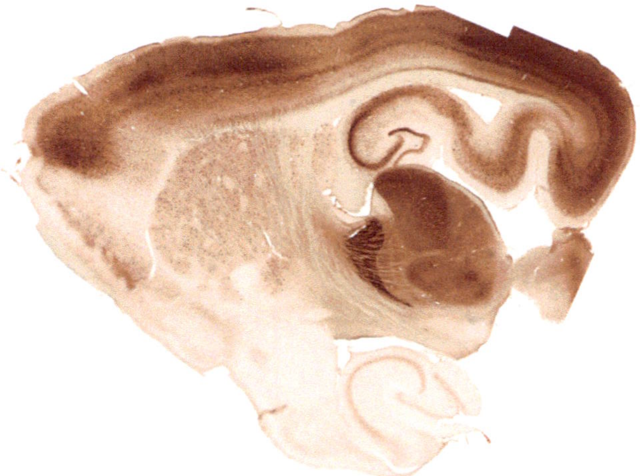

**4.18 mm**

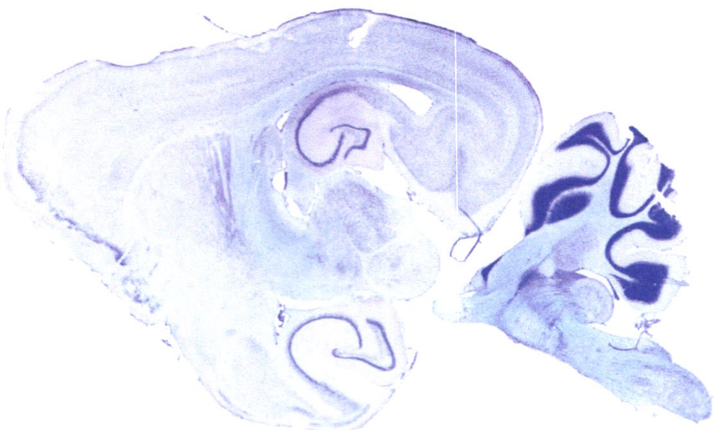

**5.03 mm**

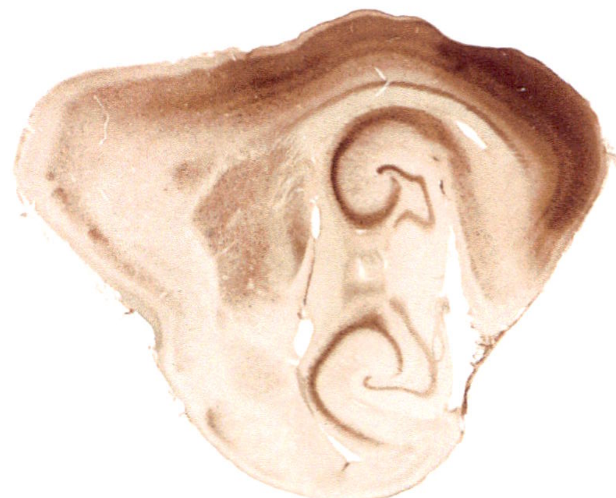

**2mm**

## Appendix 7: Sagittal sections throughout the brain – Nissl and PV staining

**5.31 mm Lateral**

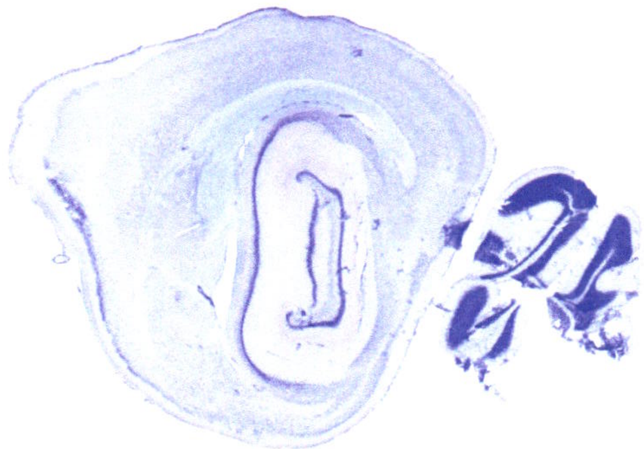

**5.60 mm**

**6.45 mm**

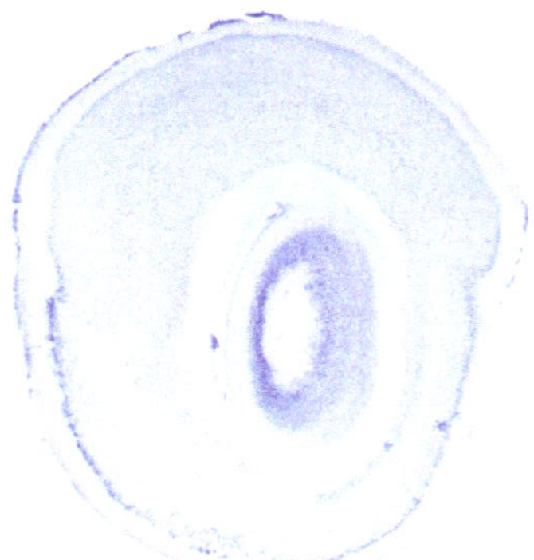

**7.02 mm**

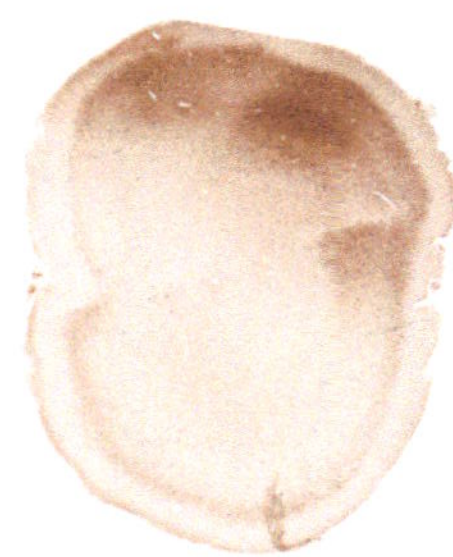

2mm

# Appendix 8

Series of horizontal sections through the brain of another bat (male), showing Nissl staining through the entire brain (first eight pages) and Parvalbumin (PV) staining of two representative sections (last page). The approximate D-V (dorso-ventral) atlas stereotaxic coordinates are indicated.

## Appendix 8: Horizontal sections throughout the brain – Nissl staining

−1.72 mm

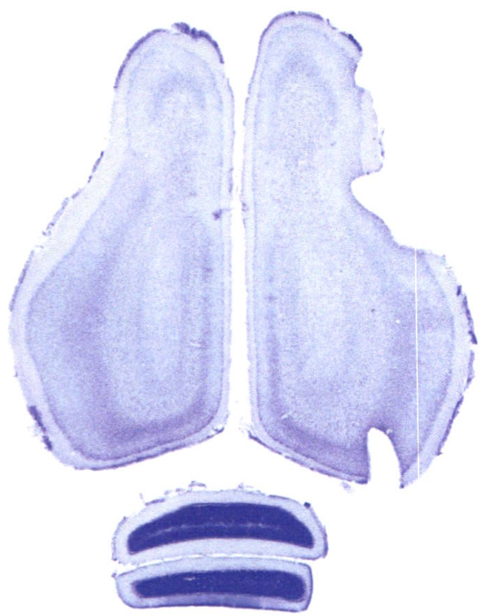

−2.28 mm

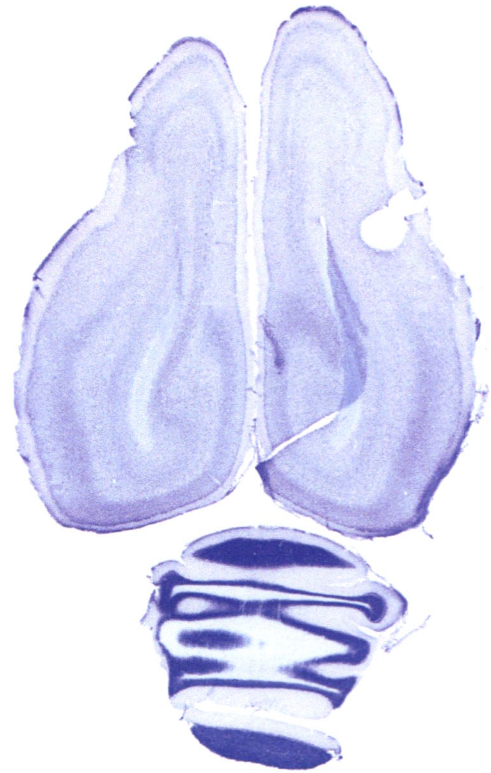

2mm

## Appendix 8: Horizontal sections throughout the brain – Nissl staining

−2.84 mm

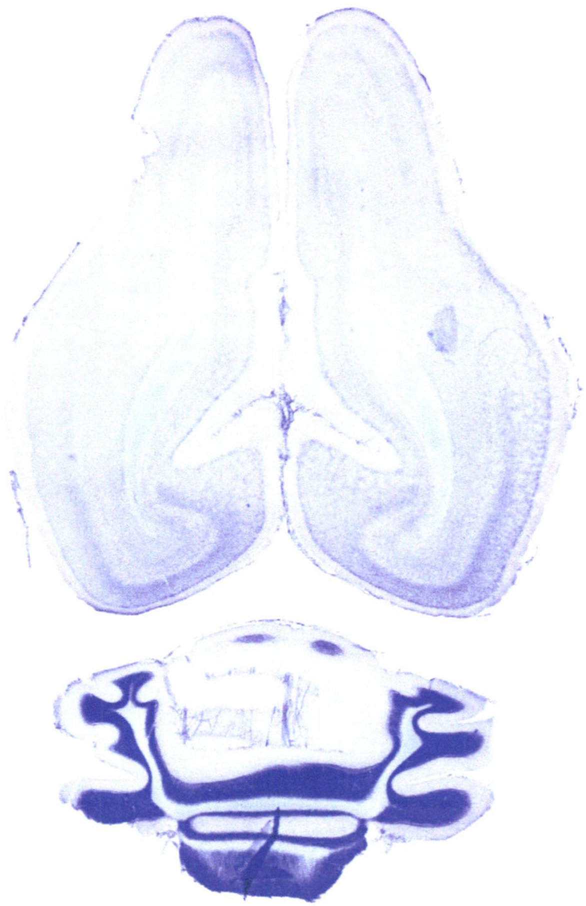

−3.40 mm

2mm

## Appendix 8: Horizontal sections throughout the brain – Nissl staining

−3.96 mm

−4.52 mm

2mm

## Appendix 8: Horizontal sections throughout the brain – Nissl staining

−5.08 mm

−5.64 mm

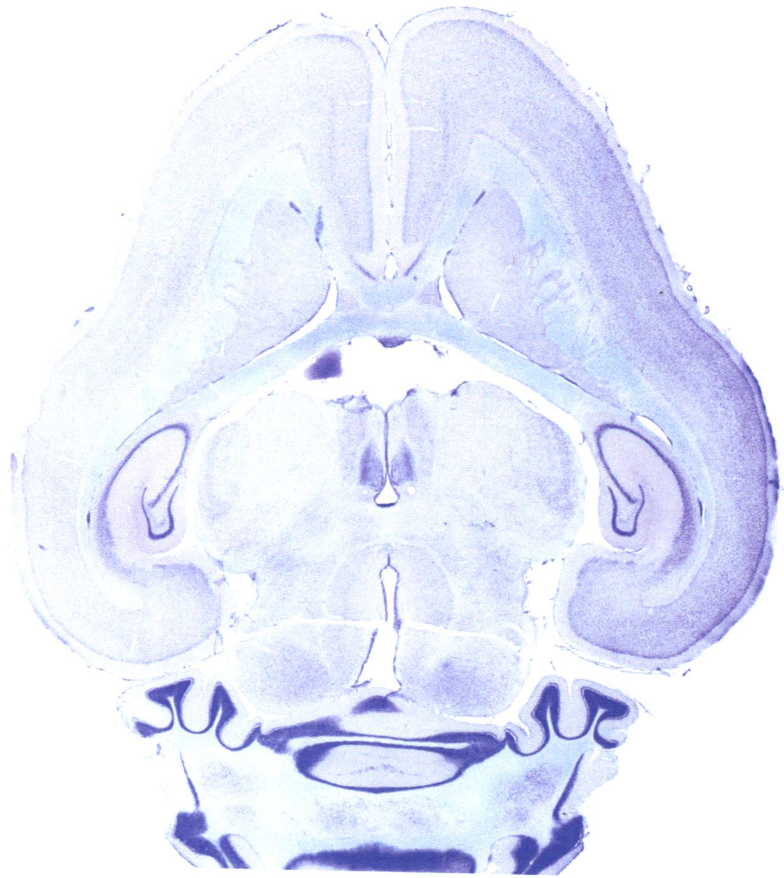

2mm

## Appendix 8: Horizontal sections throughout the brain – Nissl staining

−6.20 mm

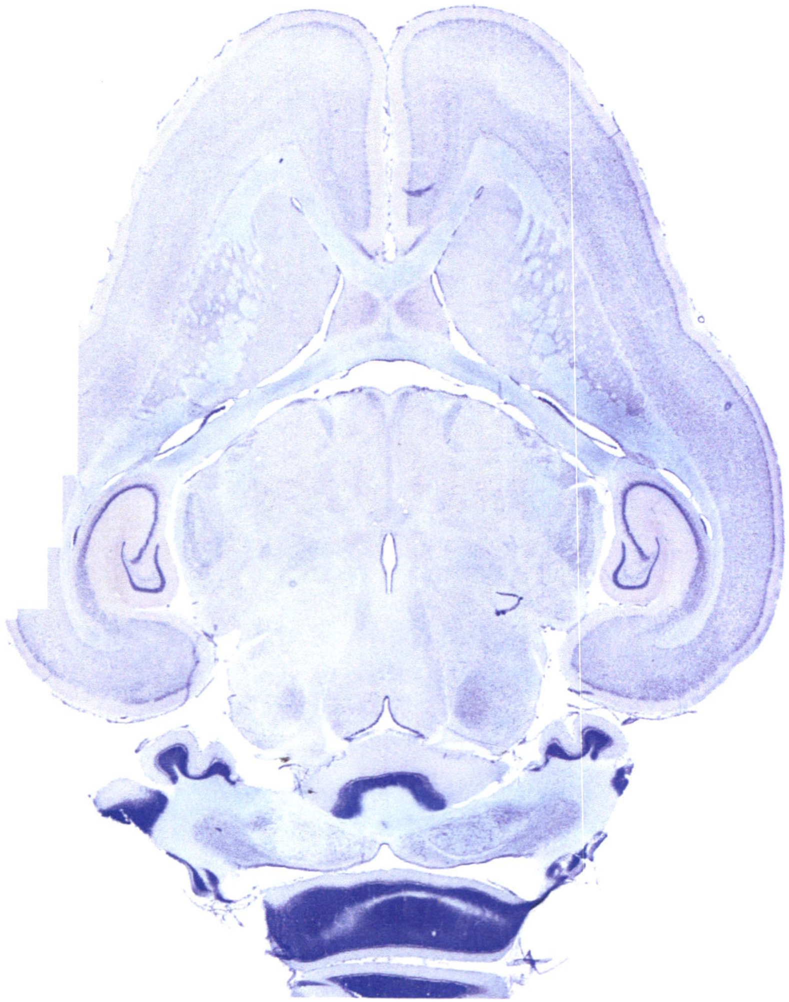

−6.76 mm

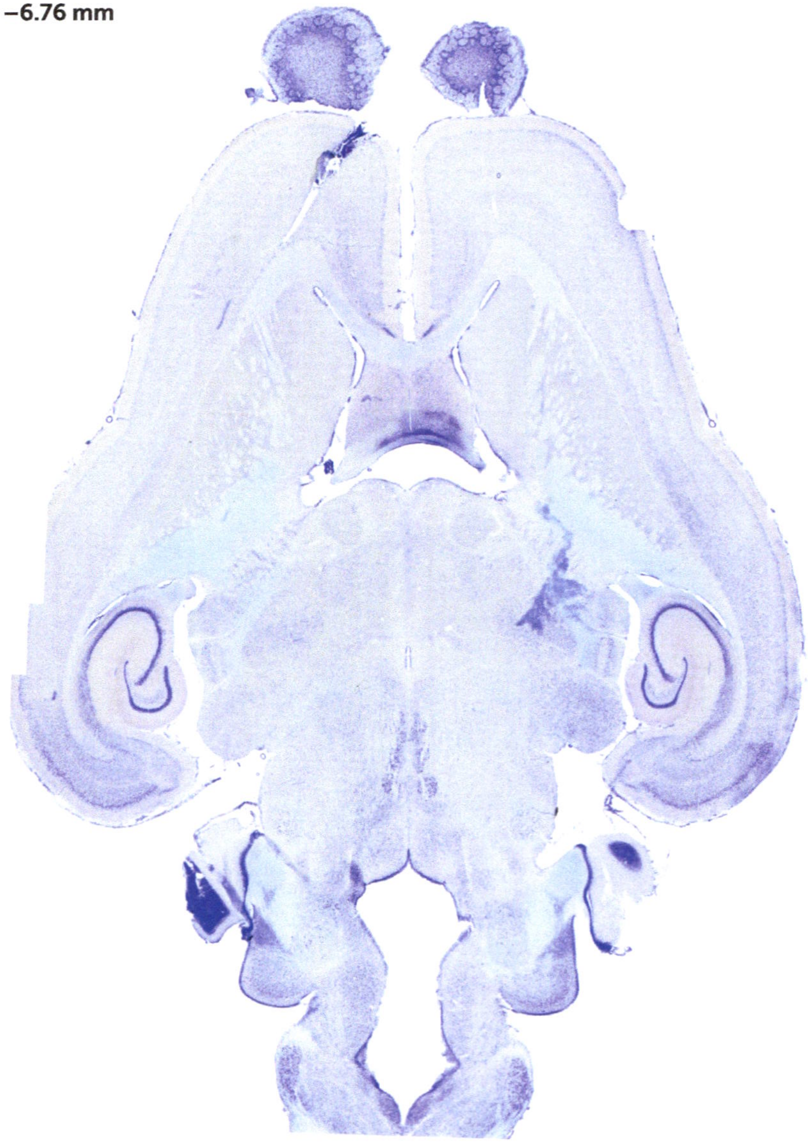

2mm

## Appendix 8: Horizontal sections throughout the brain – Nissl staining

−7.32 mm

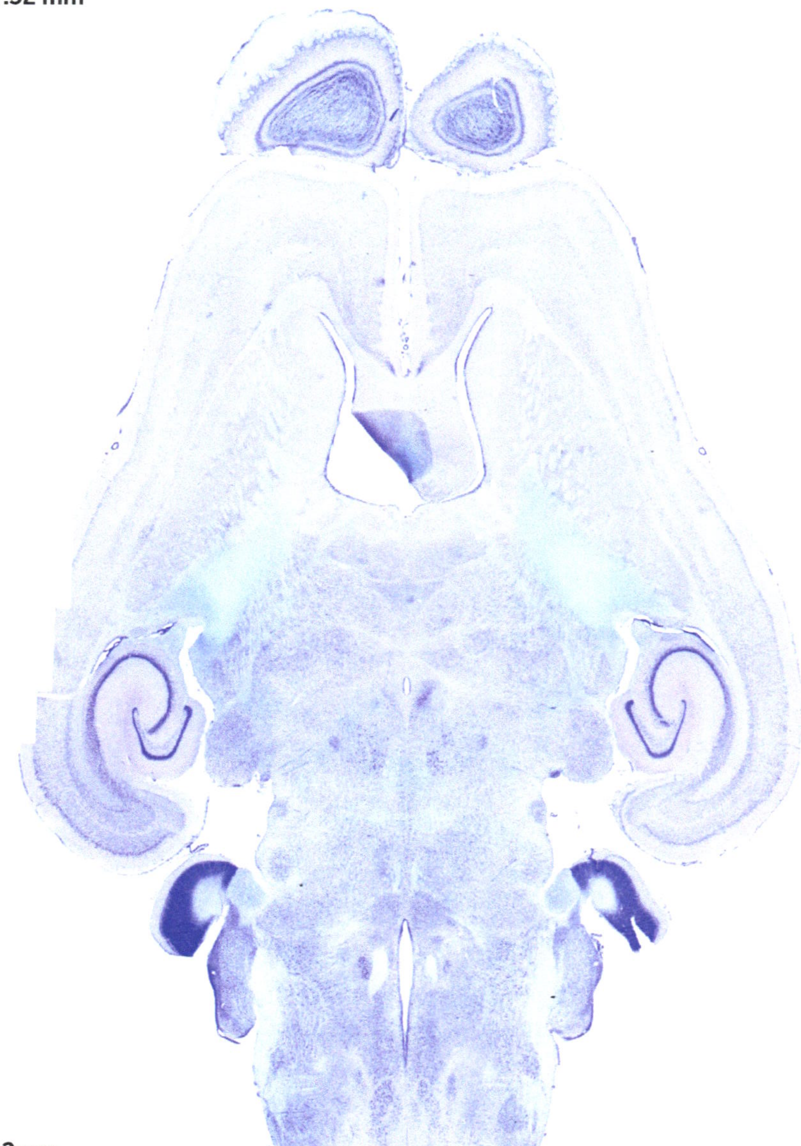

−7.88 mm

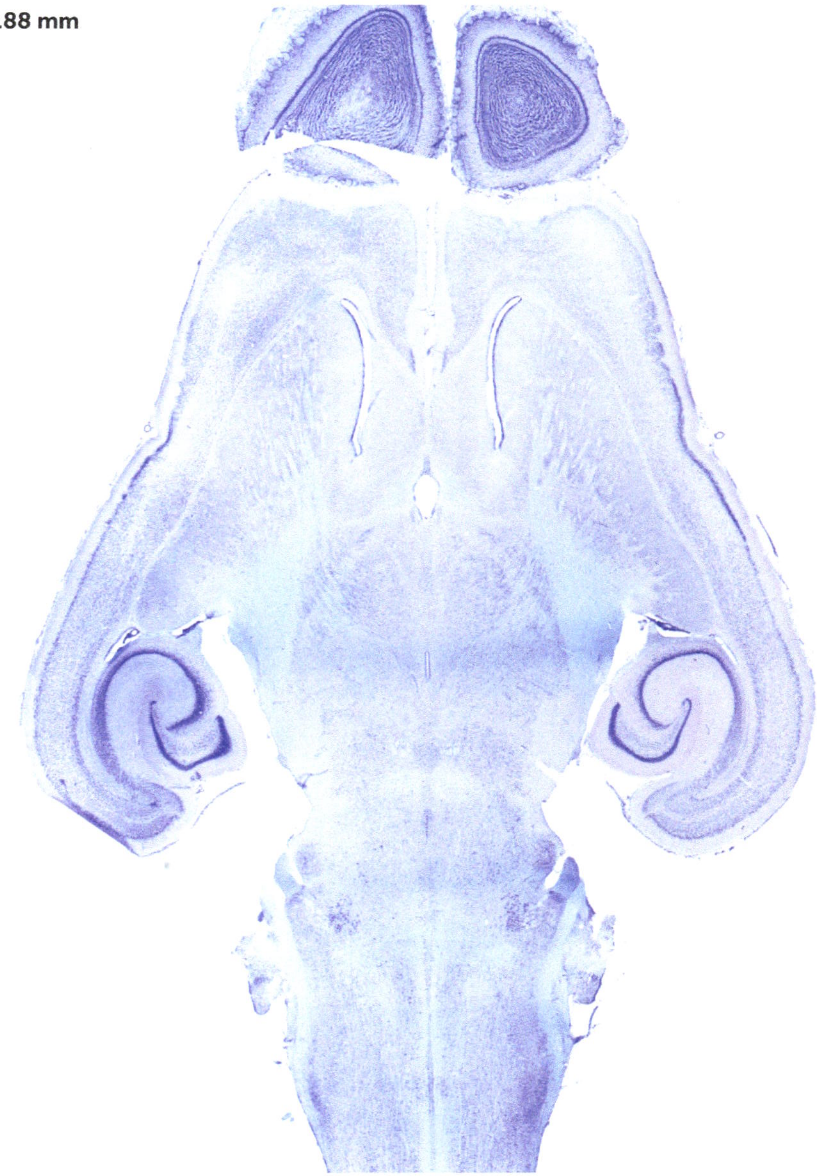

2mm

## Appendix 8: Horizontal sections throughout the brain – Nissl staining

−8.44 mm

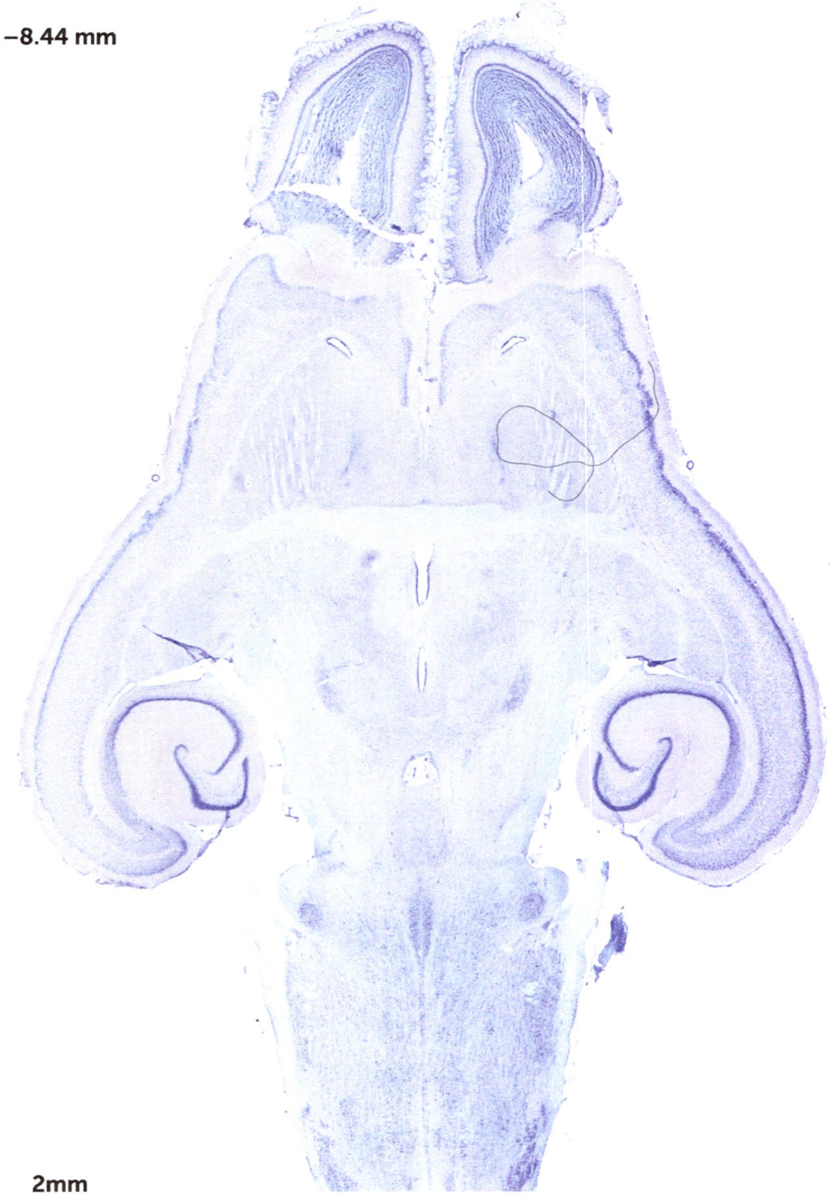

−9.00 mm

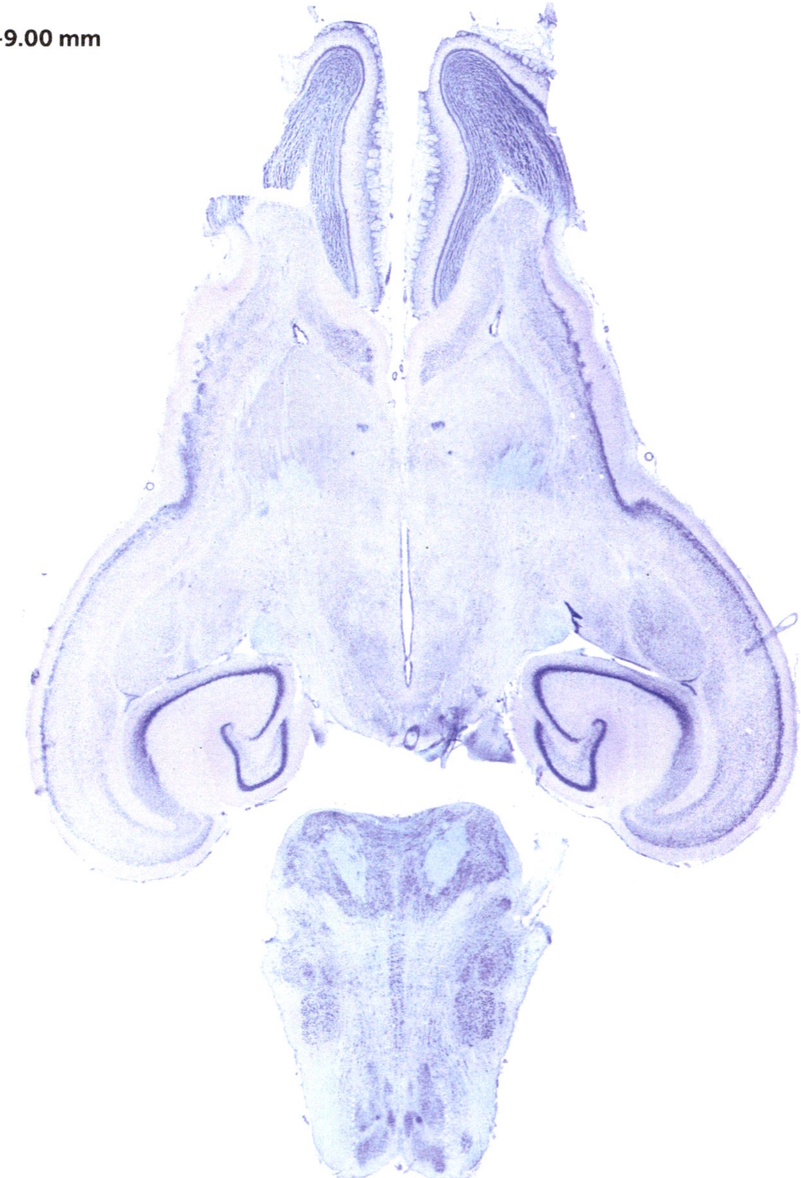

2mm

## Appendix 8: Horizontal sections throughout the brain – Nissl staining

−9.56 mm

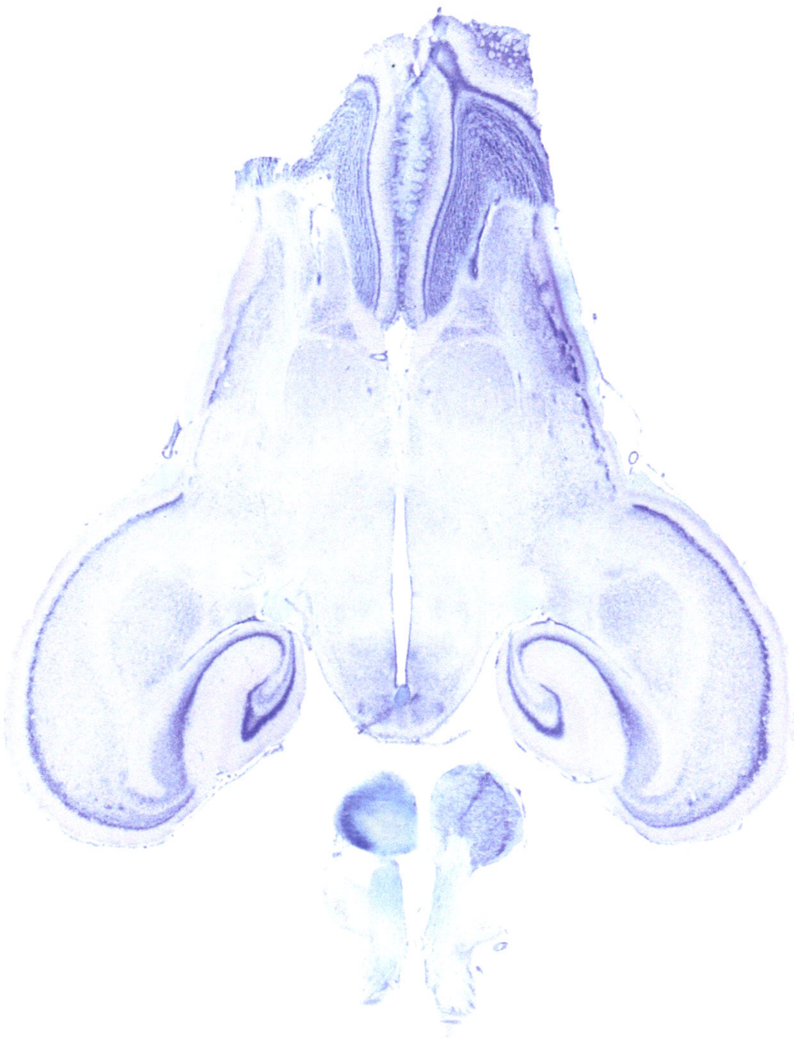

−10.12 mm

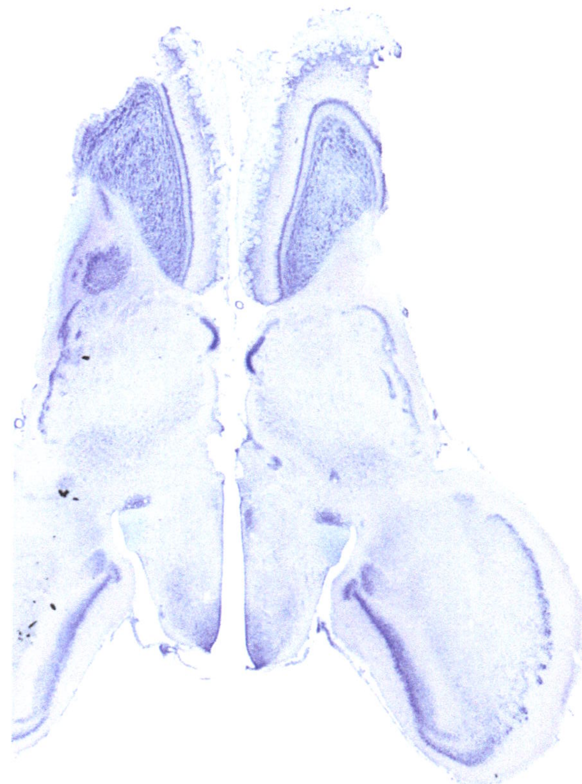

2mm

## Appendix 8: Horizontal sections through the hippocampus – PV staining

−5.92 mm

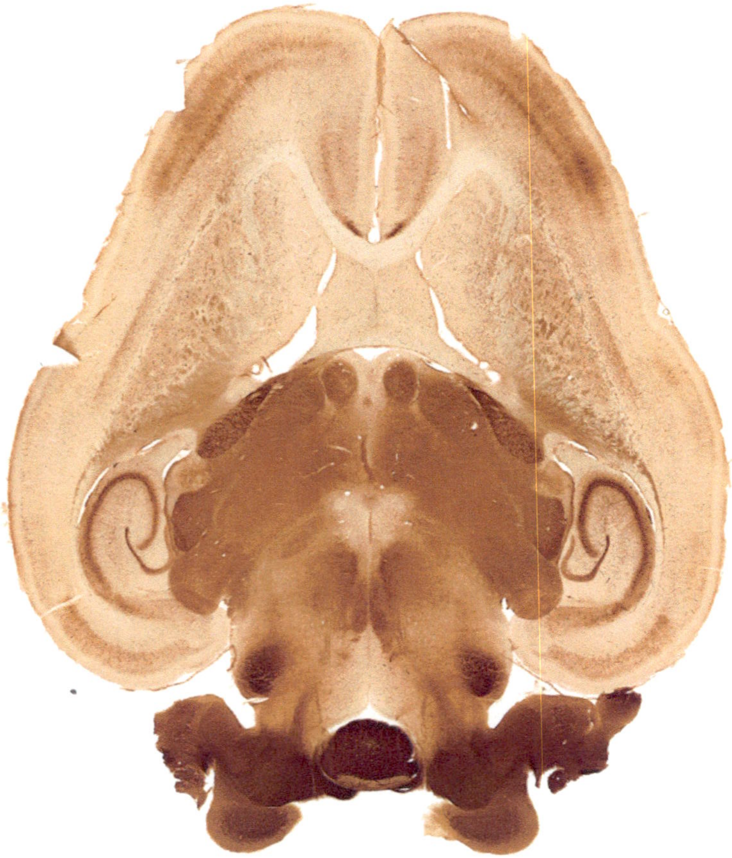

−7.32 mm

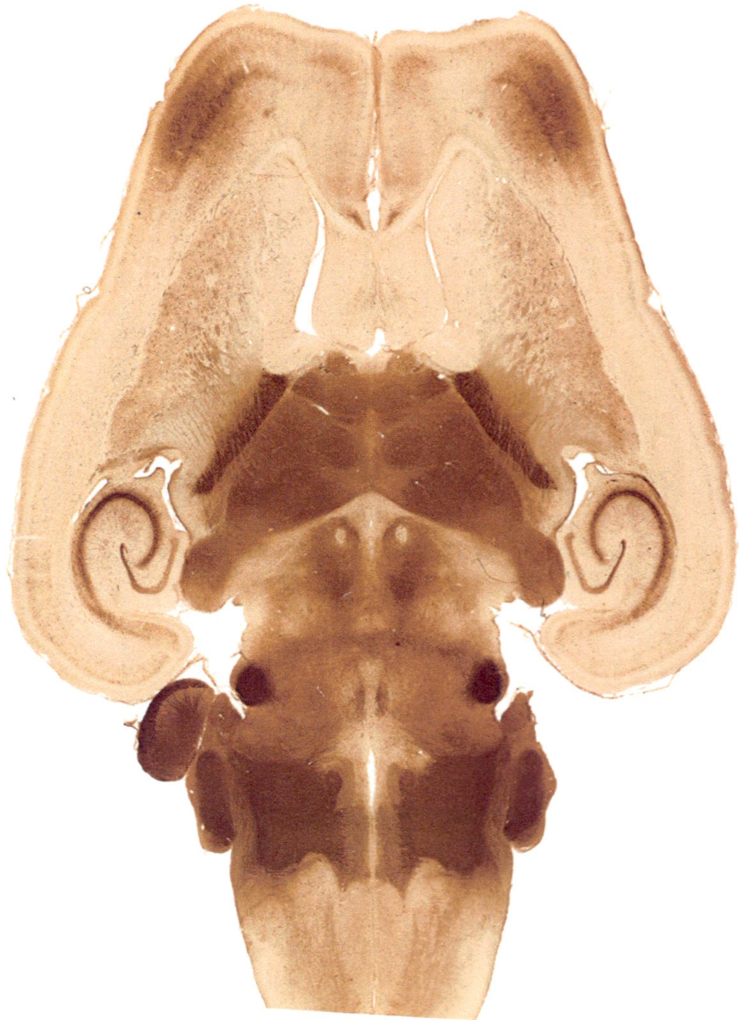

2mm